THE SALT STONES

THE SALT STONES

Seasons of a Shepherd's Life

HELEN WHYBROW

MILKWEED EDITIONS

© 2025, Text by Helen Whybrow
© 2025, Illustrations by Wren Fortunoff
All rights reserved. No portion of this book may be used or reproduced for the training of artificial intelligence. Except for brief quotations in critical articles or reviews, no part of this book may be reproduced in any manner without prior written permission from the publisher: Milkweed Editions, 1011 Washington Avenue South, Suite 300, Minneapolis, Minnesota 55415.
(800) 520-6455
milkweed.org

Published 2025 by Milkweed Editions
Printed in Canada
Cover design by Mary Austin Speaker
Cover and interior illustrations by Wren Fortunoff
Author photo by Ali Zipparo
25 26 27 28 29   5 4 3 2 1
First Edition

Library of Congress Cataloging-in-Publication Data

Names: Whybrow, Helen, author.
Title: The salt stones : Seasons of a shepherd's life / Helen Whybrow.
Description: First US edition. | Minneapolis : Milkweed Editions, 2025. | Includes bibliographical references. | Summary: "On being a shepherdess in the Green Mountains of Vermont"-- Provided by publisher.
Identifiers: LCCN 2024039883 (print) | LCCN 2024039884 (ebook) | ISBN 9781571311627 (hardcover) | ISBN 9781571317872 (ebook)
Subjects: LCSH: Whybrow, Helen. | Women shepherds--Vermont--Waitsfield--Biography. | Sheep farming--Vermont--Waitsfield. | Farm life--Vermont--Waitsfield. | LCGFT: Autobiographies.
Classification: LCC SF375.32.W49 A3 2025 (print) | LCC SF375.32.W49 (ebook) | DDC 636.3/01092 [B]--dc23/eng/20240927
LC record available at https://lccn.loc.gov/2024039883
LC ebook record available at https://lccn.loc.gov/2024039884

Milkweed Editions is committed to ecological stewardship. We strive to align our book production practices with this principle, and to reduce the impact of our operations in the environment. We are a member of the Green Press Initiative, a nonprofit coalition of publishers, manufacturers, and authors working to protect the world's endangered forests and conserve natural resources. *The Salt Stones* was printed on acid-free 100% postconsumer-waste paper by Friesens Corporation.

for my mother, Margaret Ruth Whybrow, in loving memory

CONTENTS

CHAPTER 1	The Cord	1
INTERLUDE I		11
CHAPTER 2	Initiation	15
CHAPTER 3	The Flock Is the Land	39
CHAPTER 4	Spring Meadow	61
CHAPTER 5	Stalking Coyotes	89
CHAPTER 6	Lightning	109
CHAPTER 7	Passerine	129
INTERLUDE II		145
CHAPTER 8	Shearing Day	151
CHAPTER 9	Gifts	173
CHAPTER 10	Traces	189
CHAPTER 11	Damage and Healing	203
INTERLUDE III		221
CHAPTER 12	Winter	227
EPILOGUE	A Shepherd's Mind	251

A Shepherd's Glossary 263
Notes 267
Bibliography 279
Acknowledgments 283

In the high pastures, the shepherds go find flat rocks and they line them up in the grass. These are the salt stones. Every night, the shepherds pour four or five handfuls of rough gray salt on these flat rocks. It's for the nursing ewe, it's for the trembling young lamb, it's for the good sheep huddled with cold or the one who has a thorn in its foot; it's a consolation and a remedy.

—JEAN GIONO

Once in his life a man ought to concentrate his mind upon the remembered earth, I believe. He ought to give himself up to a particular landscape in his experience, to look at it from as many angles as he can, to wonder about it, to dwell upon it.

—N. SCOTT MOMADAY

THE
SALT
STONES

CHAPTER 1

THE CORD

Under the palm of my hand, I felt the slow pulse of granite.

—JEAN GIONO, THE SERPENT OF STARS

With my fingertips in the dark, I can feel a nose and two feet. The feet are soft and pointed, positioned just above what I think is the head, only I don't know if they are back feet or front feet. I need to know because the answer tells me if I have one lamb coming out more or less okay or two lambs tangled together, with the first lamb folding its feet back and its twin doing a backward somersault over the top of it. This would be bad.

It's Sunday, about 2 a.m., I think. For a few hours I've been out in the shed that extends along the east side of our old barn, and I have that all-nighter feeling, when time is suspended and I am hovering slightly outside of my body, as if the part of me that's awake is no longer mine. Near the shed's opening, I sit in the damp hay blown over by fine snow that shines in the moonlight, and I lean my back against the cold stone of the foundation wall. Some forty-odd sheep lie around me, eyes closed, chewing their cuds rhythmically, like the circular lines of a lullaby.

A black ewe, Bluestem, lying near me on her side in an awkward position, with legs splayed, belly heaving, is the reason I'm awake. I came out to check the lambing shed several hours ago and noticed the inky-red membrane of a water bag, like the throat of a frog in full song, nudging out of her vulva. Earlier in the afternoon Blue had shown early signs: She stopped eating, separated herself from the rest of the flock, scraped at the ground to make a nest in the bedding. A ewe will do this minutes or many hours before she braces her hooves against the earth, throws her head back, and surrenders to the force of the new life inside her. Each birth attends to its own beat.

If a ewe doesn't deliver on her own a couple of hours after her water breaks, my shepherd mentor Barb told me, it's a good idea to help her. It often signals that something is wrong with the way the lamb is positioned in the birth canal. A ewe straining too long to deliver a lamb that has its head or legs bent back can end up delivering a dead lamb or, worse, rupturing her womb. I've waited to see what Blue can do on her own, but now she is exhausted, heaving on her side, her eyes wild.

Because she is such a fighter, Blue is my favorite. Four years ago I picked out ten sheep from Barb's farm in Massachusetts, including Blue's mother. The runt of triplets, whose siblings outcompeted her for her mother's milk, Blue was a wild and scrappy "bum lamb" or "bummer," stealing sips of nourishment from all the mothers in the field and only sometimes slowing down long enough to let me feed her with a bottle.

I stand up and stretch my back, wash my arm with icy iodine water from a tin bucket, and roll down my thick sleeve. I'm suddenly bone-cold. I pull my thick leather gloves over my red hands to warm them, and I step out of the shed where I can look east to the Northfield Range and collect myself. Everything is still. The stars glitter on the snow. In the ringing silence of deep winter, the sky is a black velvet curtain hung with tiny tolling bells. A few sheep have made their beds in the drifts, and they look at me placidly, their puffs of white breath wraithlike against the dark.

My husband, Peter, has taken my stepdaughter, Willow, on a trip to see old friends. My vet isn't on call this weekend, and his sub lives over an hour away. In my first lambing seasons I had mostly simple, common problems to deal with, like horn buds on a large ram lamb that made for a difficult delivery or a lamb's front leg folded back that needed straightening. Peter would help me catch the ewe and hold her head, encouraging me as I fumbled around and learned how to help, more or less by trial and error. I'd come to the barn expecting to see new lambs already up and nursing, wagging their tiny tails. I'm in new territory now. I can't help thinking about the nightmarish birthing stories that Barb shared with me: a lamb with two heads, a huge ram lamb that had to be dismembered to come out, an emergency C-section in the muck of the barn floor where the ewe was sacrificed for the lambs. *Please, no.*

A terrible moan makes me turn around. Blue is straining, her neck arched and lips in a grimace. This time she doesn't get up as I kneel beside her and place a hand on her huge belly, rigid as stone. I roll up my sleeve and reach inside her, following logic, as a caver without a light might feel her way, inch by inch along the wall to a passage that's familiar. The mass of womb and placenta are warm and wet, a jellyfish in a tropical sea, while the bony wall of the pelvis grinds against the back of my hand.

I find the unborn lamb's nose again, rounded and soft. Then, above the nose, I again feel the hard triangles of two feet. I try to lock my fingers around them to pull, but they twitch back. I've lost them! Gently, I reach farther in, find the feet, and trace the legs back against the unforgiving pelvic wall to the second joint. I do this to determine if what I have is a knee (front foot) or hock (back foot). I am 51 percent sure they are back feet. If they are, there is no anatomical way they belong to this nose, unless the lamb is a contortionist. To pull the first lamb out, I have to find its front feet, but they are folded back, out of my reach.

I step back from Blue, stand up, and rinse my arms. Her twins are jammed in the birth canal like tangled tree branches in a narrow stream during spring flood. The natural birthing position for a lamb is like a diver, head between front limbs, shoulders forward, and streamlined for the sprint to air. When the arms are back, the shoulders are too broad for the opening. This is the problem with the first lamb, I think. The second lamb has its feet over the first lamb's head, and its own head is somewhere farther back in the womb. Most likely those were its back legs I felt: a breech, or backward, birth. A breech birth can be delivered, but there's a high chance that the lamb will inhale fluid on the way out as the umbilical cord is stretched and breaks. Then it's unlikely to survive.

When our daughter, Wren, was born, our midwife lifted her to my chest, and Peter cut the cord near her belly. In my exhausted state, where images took on a strangely supernatural intensity, I remember thinking how thick the cord looked, like a sinewy tree root you find while digging, the kind that resists every effort of the spade. Something muscular and undeniable. A lamb's umbilical is as translucent and soft as a bit of milkweed down. I've never had to break or cut a cord; it always happens on its own as the lamb slips out of its watery home and onto the hay, becoming a creature of the breathing world. This astonishes me, that the cord that sustains life could be so thin. Perhaps so, being of a creature that's closer to the wild. A ewe would have to lick the birth membranes from the lamb and be on her way, to leave little trace behind for predators to smell. The lamb's ability to get to its feet and follow its mom within a few minutes of being born is an evolutionary imperative.

Would Blue die in the wild? I wonder. Probably, yes. No doubt all these thousands of years that humans have been shepherds and helped ewes give birth have tweaked the evolutionary arc so that not only the easy birthers pass on their genes. A nomadic shepherd would have helped a birth so that the whole flock could more quickly move out of the wind, get away from predators, or find the spring grass. Only in the worst cases would they have abandoned a laboring ewe and unborn lambs to the wolves.

On Vermont hill farms like ours, most of which were self-sufficient by economic necessity well into the twentieth century, another live birth would have meant hope for a family that faced spring with little left but potatoes and cabbage in the cellar. The peaks surrounding our farm tell this story—Scrag, Stark, Mount Hunger—while Shepard Brook drains their slopes to the Mad River. Sheep also meant clothing and blankets during the long period of history when wool was the primary source of all cloth, and every farm woman knew how to make homespun. Those are the practical reasons, but I know there were—and are—more important reasons shepherds would do everything to assist a birth; this ancient, primal thing of caring for a flock is ultimately about human attachment.

BLUE'S EYES ARE WEIRDLY WHITE, HER BREATH SHALLOW. SHE rests her horns against the wall and goes still. Inside her womb, I trace shapes of the yet-to-be-born with my fingertips over and over, guessing their anatomy aloud—front foot, nose, back foot—a lock picker in the dark. I have to be absolutely sure before I pull.

A faint noise comes from the doorway, where a dim light spills from the barn into the sheep shed. Wren, who is three, has navigated the dark, snowy path from house to barn in her dinosaur pajamas to find me. She comes around the corner tentatively, with a worried look, then runs quickly down the steps with arms out when she sees me crouched there. I have only glanced her way. My face, I'm sure, is a mask of concentration.

Wren climbs onto my back, and her cold hands find the warmth of my neck beneath my parka hood, her too-big boots dangling from sockless feet. Since her father was away, at bedtime I told her, "If you wake up in the night and I'm not in my bed, then look outside. If the barn lights are on, I'm out there and you can come out. I'll leave your boots by the door." I honestly wasn't sure she would figure it out, but I hadn't come up with any better options.

"Are there babies?" she whispers close to my ear.

"Yes, soon," I say. I leave Blue and put Wren on a hay bale so she can watch. I drape my huge coat around her.

I can't take her back to bed; we are in this together now. The first time she saw a birth she was two months old, strapped to my chest under my down coat as I worked to clear a lamb's airway, her tiny head so close to my hands that I was afraid of hurting her. She is old enough now to observe more closely. I wonder if I should warn her about how seeing blood can be scary, and how when a lamb is born, it's normal for it to be wet and limp, sometimes coated with a bright, sticky yellow feces that looks gross. And sometimes, the lamb is not alive.

But I say nothing. Just before Wren appeared I think I felt the missing feet that belong to this soft nose. I reach under the nose with two fingers, find the lamb's wrist joint, and unbend the folded-back legs, first one side and then the other. Then, I push the breech twin back. It will have to wait its turn. With my right hand locking the first lamb's feet together, I pull back hard while bracing my left palm on Blue's side. I hear myself groan as I pull with all my strength to get the head through the opening, and the newborn slides into the world at last.

She is tiny, white with black spots around her eyes, legs frail as icicles. The thinnest of translucent membranes covers her body and nose. She lifts her head immediately. I towel her off, wiping the birth sac from her nose and mouth, rubbing her curly wet coat vigorously to stimulate her to rise. She feels ephemeral, a ragdoll of bone and blood, water and air.

Blue responds with a soft throaty nickering that ewes make only when licking their newborns. I stand back to watch the lamb shakily rise up on her front feet, fall, rise up, fall, her nose all the time butting against mom's flank for milk. It reminds me of a sea turtle watch I went on years ago. I wanted so much to help the hatchlings to the sea, but I was told they needed to struggle into the waves and be tossed violently back up the beach again and again to get their strength to swim. This lamb is strengthening and warming with every failed attempt to stand.

As the little creature butts against my legs, trying to find a teat, and Blue stands to lick and nuzzle her, I kneel and gently help Blue one more time. I can feel the breech lamb, pointy feet, no head. His passage is open now, my job easy. But he seems impossibly long as I pull his back feet with both hands and the whole length of him, wet and shining, finally lands on the hay with a thud. His head is strangely huge, with horn buds already breaking the skin. He doesn't stir. I move fast to clear his nose and mouth, swing him like a pendulum by the back legs to shoot out the phlegm from his airway. It doesn't seem to help. No cry, no gasp. I can feel panic rising in my chest. I lay him down and kneel beside him, palpate his tiny heart with two fingers. *Come on come on come on little buddy, you can do this.* Did I wait too long? Did I pull too slowly? Aware of Wren on the bale beside me, miraculously asleep, I will myself not to cry out.

It takes only seconds for the light to die out of his eye. That unexplainable light. It was there, then it wasn't. His cord, like a snail's trace, gleams in the hay.

I think, *That was it, his whole life.*

Blue rubs her nose on him and cleans him, pawing at the ground for him to rise. Her sharp feet scrape urgently at his still-warm lifeless body, and he crumples under her hoof like a discarded strip of towel, streaks of blood and mucus staining his white fleece. Blue's nickering and pawing become louder, more desperate. It's more than I can bear. I go up the stairs to get a burlap sack, which I line with some hay from the floor. I slide him in, returning him to the dark, with the smell of clover fields he will never know. Blue heads into the night, calling loudly, searching.

In case Wren wants to see him, as proof of what death looks like—toddlers being more curious than sentimental—I decide to leave the lamb in his burlap sack by the door. Later, if she wants to, she can help me do a sky burial—an offering to the coyotes, where we lay the lamb on the rocks far out in the woods and say a blessing, knowing we might see its bones or bits of fleece scattered elsewhere in the spring. The ground is too frozen to dig.

The sky is pale yellow over the range, the stars winking out one by one. The wind has picked up just before the dawn, and I realize I can no longer feel my feet. I lift Wren, and she is warm and heavy against my chest. I know, when she wakes, the first question she will ask is about the lambs. I will tell her: "One is good. One didn't make it." And then, together, we will go see.

INTERLUDE I

Dreams are the shepherd's savings.

—JEAN GIONO, *THE SERPENT OF STARS*

The year I was pregnant with Wren, I buried four sheep from my small flock. I was a new shepherd and knew nothing. The sheep knew that I knew nothing and found creative ways to die—a drowning, a broken neck—just outside my range of vision. It was as if they preferred other realms; the veil was so thin.

The fourth sheep I found dead was Swallow, a big black ewe, mother of twins. It was late fall, the trees brittle against the sky. The ewe lay near the electric fence, part of the hot wire wrapped around her horns and one forefoot. I untangled her head, the other sheep looking on, crowding at a distance like human spectators at an accident. I took her stiff front legs in one hand and back legs in the other and tried dragging her across the grass, away from the woods. I couldn't dig there, through all the tree roots. I was five months pregnant and not at all large, but I felt deliberate and slow. Her head lolled back at a terrible angle, but I couldn't carry her.

Steam rose from Swallow's body as the sun hit the frost covering her fleece. In the hollow where she died, earthworms had come to the surface to be nourished by the warmth and moisture of her fleece. Dragging her made a silver trace through the grass. Softly, it began to rain. I went to get a shovel, then rested on a rock in the gentle rain before digging a hole large enough to hold her.

    All night the rains came down. I dreamed of birthing a child, my child. I was at home, and another woman was there, a woman whose face I never saw but whose hands and voice were steady and warm. I felt the contractions high up in my belly and could see the baby turn and begin to swim, following the river out of my womb. I could see the baby through her whole birth, as if my birth canal were transparent, or as if I were inside it, swimming with her. There was a place where the river channel was narrow, and I heard the woman say, "Push now," and with two strong pushes, out the baby swam. She was wet and slippery, with a seal's silky pelt. Her fur was gray with faint silver stripes, and on her head were two budding horns, like a lamb. I wasn't concerned about the horns and thought that once in a while a human baby must be born with them and then lose them shortly after.

    I saw the umbilical cord attaching our bellies, long and ropy and bright red. "Where's Peter?" I asked in the dream. "He's supposed to cut the cord. He wanted to." I knew he was somewhere nearby, doing something that needed to be done, and I knew he wouldn't have expected the birth to be so quick and easy. I was sad that he wasn't there to be part of such a beautiful thing. The midwife put the soft animal body of the child up to my chest, and she began to nuzzle for milk. I was surprised, as her mouth latched around my breast, how gentle it felt. She no longer looked like a lamb but like a human baby with dark hair and black eyes. She lay in a basket in the living room. There were lots of people filling the house with talking and laughter—my father, mother, and sister, Peter, and others—and the baby was like the stone of gravity we all came toward in our wanderings, pausing to look down on her as she slept.

I WOKE FROM THE DREAM AND REACHED ACROSS THE BED TO feel Peter beside me. He had come home very late that night after being away. In the shower he stood behind me, soaping my belly. There was too much to say for words. You can want a child more than anything in the world. The thing people don't tell you, that's too big to talk about, is the fear that comes the moment you realize—the moment you dig a hole in the rain—that if anything ever happens to this child, you cannot imagine how you will possibly survive. There isn't another love like this. It's beautiful and terrible.

When the rain stopped I went down to pile stones on Swallow's grave. I moved a few flat stones from the old wall and stacked them. The four graves made a circle around the fields.

Seemingly from nothing spring these vigorous, beautiful lives. And then, so soon after, they become nothing again—and yet also part of everything. We all come from the earth and return to the earth, and all our days between are nourished by the earth and her soil and rain and sun and fruit. Some lives—a lamb, a dragonfly—are so brief. In themselves they seem to have little purpose. How do we look on such a life and make sense of it? Or is there none to be made? Is it enough to know that each life in its creation is essential, part of earth's regeneration and abundance, and that each death is also an essential and equal part of earth's fecundity?

As my friend Ellen says, "No, it is not enough, but it is something."

The next night we tell Willow—before bed, with candles lit—that she will be a big sister soon. We tell her that our midwife, Laura, will help deliver the baby at home. She is very excited, wants the baby to come now, and says that she is going to stay awake until it comes. And she also says the baby better not grow up too fast, because she always wants to be the biggest. She feels my belly—I tell her the baby looks like a big pollywog in the pond, swimming around. She puts her ear against my skin and listens intently. "Better not spill your tea, or the baby will come out!" she declares. I laugh. Her five-year-old logic puts me at ease. I begin to surrender to the mystery.

CHAPTER 2

INITIATION

> I went along and my thoughts, like a fledgling bird, learned to fly, too.
>
> —JEAN GIONO, *THE SERPENT OF STARS*

I SEE A SMALL SQUARE PAPERBACK, SLENDER AND UNASSUMING, on a cluttered table at the Tempest Book Shop in our little town. Wren is two weeks old and asleep in a pouch under my winter coat, sweaty cheek pressed against my breastbone, eyelids like blue eggshell.

The Tempest is where Rick hangs out behind towers of boxes, gives away fragrant loaves of homemade sourdough bread every Saturday, repairs watches, and tinkers with model trains that clank around in the ceiling. If you can get through the aisles overburdened by unpacked boxes and stacks of musty used books in no apparent order, sometimes you find a book you're looking for. Or, more likely, one you're not, like this one, *The Serpent of Stars*.

I had stopped in for bread. I wasn't looking for a book at all. I don't have any time to read. At the publishing imprint where I worked in my former life, it was my job to read: manuscripts, new books, reviews, magazines, looking for new authors. Since I started farming and shortly thereafter had a baby, I can only get through a paragraph or two before falling asleep. I miss a life in books the way you miss an old lover. You can't go back, but the nostalgia is a kind of ache.

Huge, buttery, mouthwatering loaves of sesame-sourdough, cocoa-cornmeal-molasses, and whole wheat-walnut are piled on a table in brown paper bags. They lie on top of a once-curated but now-abandoned display of books on natural history and the land. I am distracted by this small book with an uncoated blue paper cover. A faded photo of fading light on a mountainside. A man in a long black cape and beret standing in the mist, facing away, watching. I can just make out a flock of sheep like pale standing stones.

I sway slightly from side to side as I look at the book with its intriguing title, one palm cupping my baby's rounded form. This is my first outing with her, and I fear that at any minute she will wake up and start crying. I have been in the cocoon of those first sweet, scary days of motherhood, and I don't feel like talking to anyone yet. Rick, who is looking over his glasses, in slippers, and with an untied bow tie around his neck, will tell stories for hours, so I hastily leave a ten-dollar bill on the counter to cover the book plus tax and duck out. He won't mind. I can hear him crashing around behind the register, looking for a book of poetry he has just recommended to a frightened-looking customer who said she needed a gift. It might take a while.

In the car I miraculously transfer Wren to the car seat, still asleep. Since the night she was born in the farmhouse fifteen days ago, her body has mostly been against mine—nursing or sleeping on my chest—and strapping her tiny form into the car seat makes me feel hollow and anxious. Why have I bought this book and forgotten bread? I don't dare leave Wren to go back inside. I flip the slender book over. There is no jacket copy, just a quoted passage from the story: *Open yourself!* it reads in italic font. *Here you are crossed by the suns and the clouds; here you are traveled by wind. Listen to the beautiful wind that dances over your blood as over*

INITIATION

mountain lakes; listen to the way it makes the beautiful sound of its depths ring out!
I begin to turn the pages. I know this author, Jean Giono—Peter recently read his novel *Joy of Man's Desiring* and loved it—but this book is newly translated from French by Jody Gladding. For a couple of years after college I worked for a small Vermont book publisher whose first success was a translation of Giono's *The Man Who Planted Trees*. I became friends with book artist Michael McCurdy, who made the exquisite, intricate woodcuts to accompany Giono's simple story about a shepherd who transformed the blasted hillsides of southern France after the First World War by planting acorns. Here in my hands was a Giono novel about the same length as the one I knew so well but written in an utterly different register—full of mystery and ecstasy and simmering violence, a Whitmanesque stream of consciousness mostly told at night around a fire, where shadowy figures perform a shepherd's play.

THE SERPENT OF STARS, ORIGINALLY PUBLISHED IN 1933 AS *Le serpent d'étoiles*, begins with shepherds, or rather the lack of them. They are missing at the marketplace, where the narrator begins the book, and he goes in search of them. He finds one, who becomes a gracious if enigmatic host, traveling across the Mallefougasse plateau. Much like the character in *The Man Who Planted Trees* (who is also a shepherd), this host is a paternal figure, knowing things, living in a remote high place, in tune with the land as an extension of his being. But the land is the main character. The wind, the hills, the night, are all creaturely, animate, and insistent, both threatening and life-giving. Everything is connected and yet unknowable.

Early in the journey, the narrator and shepherds make a meal of grass and night: "The night we munched with the salad. The night overflowed from the crater in slow gushes, and our mouths were full of night when we bit into the bread crusts rubbed with garlic." The grass and night, bread and oil mixed in the belly and were breathed out again into the "saps and smells" of the terrain, "so much so that finally, we were drunk from the triple power of the sky, the earth, and truth."

Nine shepherds come to a place deep in the mountains where they perform a play on the summer solstice. They wear red scarves around their heads. They light a fire. They take elemental roles like the River, the Mountain, the Sky, the Cold, the Rain, the Beast, the Man. An ocean of sheep surrounds them, breathing softly in the dark. They tell stories that rise unbidden from that woolly darkness.

I READ THE BOOK IN ONE GULP SITTING IN THE CAR IN THE Tempest parking lot, my baby asleep, my blood on fire. The words cut directly to my longing in the tightly wrapped cocoon—enraptured, but also in bondage—that comes with birth. I was beholden to a baby and a flock of sheep that was almost in lambing, to things that needed my body to be present day and night. His words found the ragged nerve endings of new motherhood and began, in ways that I didn't yet understand, to connect the mundane in my life to the mystic. Sometimes it takes a fiction or a dream to fill the place in your heart that you didn't even know was searching. The play made no sense to me at first. I read a page and reread it, trying to make sense of the words. Some things you understand immediately, and others—important things—you understand over time. Like many initiations, this one settled in stealthily, like snow in a dense forest of fir.

All I knew at that moment was that I could see myself roaming the mountains with these shepherds, and that image awakened something in me. I was a young mother, a new shepherd, and I had just moved to a two-hundred-acre farm. I didn't yet know what I was doing in this new life. Just a year before I had had a singular sense of freedom, and now I was responsible for so much life that required me to be bound to one place. This book pulled at my imagination, and in time its depiction of life with a flock of sheep appeared to me as a bridge between my opposing inner worlds of longing for elsewhere and at the same time wanting to belong to something that needed me. The story was also a bridge between words—poetry—and the land, which have always been my two primal loves. Wandering with their flock, these shepherds and their mythic play were full participants in a world as vast and unknowable as the sea. They were utterly part of something larger than they could hold, and they were creating meaning from that mystery. They had *immersed*. I wanted the same for myself.

I GREW UP ON A SMALL FAMILY FARM IN NEW HAMPSHIRE, IN a fold of damp earth in ancient hills that slope down to the Connecticut River. Like where I live now in Vermont, it is Abenaki homeland, taken in the early eighteenth century by wealthy farmers who came steadily upriver from southern New England searching for more open land. Our farmhouse was built in 1777 by the first generation of white settlers, and the woods were full of old stone foundations, walls, and unmarked graves. Nearby was the Cornish Artists' Colony, where Augustus Saint-Gaudens, Maxfield Parrish, and other famous artists built elegant homes in the early 1800s. Next door to us throughout the 1970s and '80s was a commune organized by Donella and Dennis Meadows, systems thinkers and environmentalists who published, when I was four, the best-selling book *The Limits to Growth*, predicting the end of the earth's resources—a reality that no one questions now but was a radical idea at the time. My parents found the land they fell in love with next door to the enclave of artists and intellectuals, nature lovers, and gardeners that the Meadowses cultivated.

My parents grew up in working-class homes outside of London during World War II, fell in love at fourteen, and at sixteen rode a tandem from London to Cornwall. They came to the United States on the *Queen Elizabeth II*, when my sister and I were babies. Though my grandparents had all left school in their teens in order to work, they deeply valued education for their children—my mother graduated with a degree in social work from the London School of Economics, and my father became a doctor. Most of what my parents brought from England were books, including a musty, leather-bound copy of *Nature Through the Year*, whose stories of the handwork and natural rhythms of the English countryside depict my father's upbringing and had a deep influence on my young heart.

Our homestead farm in New Hampshire was something of an oddity in a village full of generations-old dairy farms. My parents weren't farmers, but they had grown up frugal and practical, and they were adventurous and full of energy. My father designed a saltbox barn and tractor sheds and filled them with ancient haying equipment from farm auctions; my mother planted a huge garden and helped run the local 4-H club, of which my sister and I were die-hard members. We kept Jersey cows for milk, selling the surplus raw to our neighbors, and my mother put up jars of tomatoes and pickles and filled the root cellar and freezers with beans, corn, squash, and potatoes. We had asparagus in the spring and apples in the fall and even made our own butter and fresh farmer's cheese. As kids we would hawk beans and zucchini for pennies at the end of our driveway to anyone who drove by, which was basically no one. Best of all, we took our cows to the fair in the summer to show them for 4-H, lying down to sleep at night in the sawdust next to their warm, heavy bodies while the sounds and lights of the rides swirled overhead and the night frost coated the cows' wet flanks like sugar.

I have one older sister, Kate. We were in charge of all the barn chores, morning and night, seven days a week, and we divided and rotated them equally in a complex ritual known only to us: who on which day of the week scooped the fragrant, sweet feed from the giant oak whiskey barrels into the wooden bins; who ran the shovel along the concrete channel full of cow shit and shavings to wheel it out and push like mad up the ramp to the top of the manure pile; who squatted by Cleo's or Hazel's side, tied her legs, and cleaned her teats for the milking machine, remembering to squirt some steaming milk into a dish for the many swishing cats; who pumped the water from the rusty red pump or held a bottle for the calves. Who went out, last of all, to collect the chicken eggs, which would smash in my pockets as I instantly forgot about them and belly flopped onto my sled for the swoosh down the hill to the glowing house in the frigid winter dark.

For Kate and for me, all through childhood, the farm was always the biggest and best thing in our lives. But by the time I contemplated college my enduring passion had come to feel like a shackle. It was a shackle in practical ways—the cows needed milking morning and night, a ritual I had participated in daily from age eight—but also in emotional ones. The farm, as aware as I was of my love for it and my great privilege to have grown up there, was a polished stone around my neck, the kind your fingers reach for without even thinking, the kind that soothes but also hangs a bit heavy, whose strings you sometimes wish you could untie. I was so attached to the place of my childhood that my wanderings in my late teens and early twenties were haunted by intense, debilitating homesickness.

I was restless and full of longing, and every message that seeped through my pores as an eighteen-year-old in America told me that losing the attachment to one's childhood home was necessary to finding oneself, to differentiating from one's parents, to growing up. People in my village would shake their heads at the young people who had chosen to stay home and not go to college, as if it were a terminal illness, a great shame. Only those with no ambition or those who had suffered great trauma stayed home. And yet I also saw how the old farming couples had no one to help them and that their farms were dying. Our neighbors, who were salt-of-the-earth people that you never saw indoors and who rarely went as far as the next county, sold their dairy farm to a developer and moved to Florida. It seemed desperate, not the way things should go.

I couldn't wrap my head around this contradiction. What was happening in my own heart was happening on a cultural scale. How was I to reconcile these dominant cultural messages with my own abiding love for the family farm, this sense that to leave was to sever a limb? I was embarrassed even to talk about this devotion, as if my attachment to place were equivalent to the inability to give up a raggedy childhood doll.

The first time I left home, traveling somewhat timidly through the West, Mexico, and France in a gap year before college, I discovered the seventeenth-century Japanese haiku master Matsuo Bashō and his beautiful book *The Narrow Road to the Deep North*. I was struck by the image of his grass hut under the banana tree, which he left one day to walk into the mountains, an old man whose small satchel cut into his thin shoulders and whose path led in one direction with no thought of return. As Jane Hirshfield wrote of the poet, "Basho's haiku are the record of what the world placed in the open begging bowl of his perceptions."

I longed for that kind of freedom, the ecstasy of renouncing all one had, the begging bowl of one's perceptions being empty to receive. My bowl had been so full, my idea of self so tied down. My longing came out in the simple poems I wrote then, in lines like, "leaving home, I leave the door open to swing in the wind, with a tree losing its leaves." I wrote romantic haiku: Dipping gourds at night / In the cool lake / We swallow the moon.

In my twenties I traveled with more relish, sometimes living out of a backpack for months at a time as an outdoor instructor, between bouts of publishing jobs that kept me too much indoors. But my attachment to the farm was always there. The land where I grew up and my unconscious were no longer distinguishable to me, and perhaps never had been. It was not a general homesickness I felt, the simple craving for a place where things could be pulled out of the backpack and put away for a while. I lived with a deep longing for *one* place, a place that I felt needed me as much as I needed it. My parents, inexplicably to Kate and to me, had split up during my senior year in high school, and I worried desperately about what would happen to our home. My father's career in medicine had pulled him away to Philadelphia, and my mother stayed behind, but the place was too much for one. The farm's loneliness in those years was my loneliness, my emptiness very much its own.

I have read that there is a stage in a child's early development of language when words are primary and fixed: *Fire* is the certain flare of heat and fear the moment you understood the word for the first time; *kettle* is the one kettle with a particular shape and color that your parent pointed to and said the word. Perhaps, even after we've grown and learned to abstract language, the images of those first words, if not the words themselves, make up our psyches forever. Not just words, I think, but the first imprint of sounds and smells also stay with us as something etched with a particular time and place. We can leave, but we take those imprints with us, deeper than memory. Whenever I hear a crow in the distance, I have a flash of the smell of rotting leaves and an image of rivulets of sand moving under water at the side of the driveway of that old house. Why a crow and the spring thaw would be linked, I don't know. Perhaps I was five or six, crouched down and building a dam in the ditch by our farm road, when I first imprinted, utterly, the sound of a crow's voice.

No matter where I set up camp in the decade after my parents' divorce, in a college dorm or a tent, it was the farm—which my father named Blow-Me-Down after the brook that ran through its fields—that I would return to for solace and balance, sometimes living there for long periods in a respectful intimacy with my mother, who inhabited the house in her graceful way around the edges and ghosts of my father's things. I couldn't imagine home to be anywhere else. There was no separating them for me: landscape, identity, nature, emotion. Without naming what I was doing then, I was making important pilgrimages—journeys of the heart—back to my ever-familiar home, pulling the old stories of loss and love around me like a blanket. My soul had sealed itself around that land like a tree root around a rock. It held me up; it was my salt.

INITIATION

ANOTHER CHAPTER BEGAN WHEN MY MOTHER PUT DOWN the torch she had carried for the family for a decade, moving away to share a house with her new partner in a nearby town, the same town where she had built a community health center and greenhouse for folks with chronic mental illness to grow a nursery business. I was living in Vermont, running a small book publishing imprint for a larger firm in New York, and I would drive back home from time to time. One October weekend I decided to go to the farm to cut back the perennials in the front flower garden for winter. The day was still, with a transparent light that can come in autumn, the only time the land looks like a Maxfield Parrish painting, all blue and gold. (Parrish painted the same view of Mount Ascutney that we looked out on from our fields.)

The farmhouse was empty by then and beginning to show signs of neglect. Maple leaves from the gnarled trees in the garden piled up around the back door, swirling in eddies and rustling on the slate. Caught at the door latch by a momentary shiver of fear, I decided to turn around and walk through the tall grass up to the barn. A pair of mourning doves sidled a few steps along the sagging garage roof as I passed, as if making room for me. I rolled back the big red barn door, half expecting to hear the sudden creak of stanchion iron as the cows turned their heads toward me, and to see our horse Cricket reaching her black velvety lips for an apple. But it was empty, very still. Dust and chaff, like tiny insects, suspended in a stream of light. Onion skins, from a neighbor who had cured the harvest there, littered the floor of one of the horse stalls and the entire length of concrete where the cows once stood. Rats had chewed through the back wall of the barn where the wood was soft and mossy from the splash of rain.

The place needed me. And so, I moved back. I thought I might stay forever. I thought of N. Scott Momaday's Angela in *House Made of Dawn*: "She would see into the windows and the doors, and she would know the arrangement of her days and hours in the upstairs and down, and they would be for her the proof of her being and having been." And they would be proof, too, of our family's being and having been. The way I remembered. It was not just love and identity that drew me; it was wanting to preserve something of what we all had. My father talked about returning one day and about keeping the house for family gatherings. He, too, was memory's hostage. Can we outlive the stories of loss with enough time or rewrite them so that they no longer have the power to undo us? This was something I wanted to know.

In moving back, I learned from and respected my mother more deeply than ever before. Caring for that land and old house alone—all the ancient pipes, a dirt cellar that flooded in spring, gardens lilting toward wildness and rooflines toward gravity—one had little time for nostalgia or regret. Necessity cleaved a clean split between wishes and truth. Like burying the horse in winter, one had to get along with change. I understood, too, how hard it had been for her to live there reknitting the stitches that had unraveled, that still bore the crimp of their original shape. She had held it together for so long, and she was finally ready to let go.

My two years at Plainfield, caring for the land alone, were healing years. They were also a test of questions I'd had for a long time: Are attachment and responsibility linked, and is one hollow without the other? Can one make a meaningful pilgrimage home if one has not grappled with what it feels like to leave? Most of all, can you lose or let go of something beloved and still carry with you all the sense of belonging it gave you?

Little did I know that a still greater test of these questions would soon present itself to me. Two years after moving back to the farm in my late twenties, I fell in love in a different way. Peter was the wingbeat that flew me out. Though I barely knew him, I was drawn to this intense man full of compelling visions for his life that he was ready to follow without including anyone else, not because he didn't want you to be there, not because he wasn't generous and kind—he was both of these things—but because he was on a mission, and actions to him spoke louder than words. He required a yes or a no, never a maybe. He was definitive, and I had been living in a world of questions. If he asked me, I would have followed him anywhere. And he did.

Ironically, I followed Peter across the Connecticut River and about eighty miles west, to a farm he had fallen in love with, a farm that would turn out to have at least as much history as the one I left. The first time I saw the land where we now live, I was trespassing on a trail of blind love. The fast-walking, handsome man I followed out of the parked station wagon, up a rutted logging road, and across a late-winter crust of snow into a field was as much a mystery to me as the land he had taken me to see.

Peter and I connected while working on an anthology about land conservation for the Trust for Public Land, where Peter had worked for a decade. He hired me to edit and produce the book. The anthology marked the first sparks of change in Peter's career, when he began to publicly question the privileged underpinnings of land conservation and articulate his vision for a broader movement that could no longer shy away from addressing equity, dispossession, and social healing in its work of protecting land. "Who does this serve?" Peter began to ask in all aspects of the work, holding the privileged, white conservation movement's feet to the fire. He had begun to dream of a place where very diverse groups of leaders could come together to ignite a fundamental shift in the movement, opening hard and essential questions that come from the dark history of land in this country and that could start to change patterns of power. The place where Peter dreamed of creating this "refuge for land and people" arrived one day in the form of a flyer from a land trust in Vermont, inviting proposals for someone to steward a two-hundred-acre farm. Without knowing much about it at all, just seeing the photo of the land on the flyer, Peter had a strong intuition that this was where he was meant to do his work.

"Come with me to see this place," he said. "I haven't told them we're coming, but let's just go have a look."

It was early spring. We walked into a high meadow through clumps of white birch and a grove of massive white pines. I was wearing running shoes, and they sank through the snow's crust, sharp, cold crystals grinding against my ankles. The wind on the hill to our north sounded like a distant train bearing down on us through the trees. Beyond the pines the land opened to reveal rounded mountains in three directions, snow-covered and indifferent, the sky weighted with ice, and far below, a river flashed its silver belly as it curled through a meadow into the forest. We could just catch a glimpse of an old farmhouse down the hill, dwarfed by a barn and a massive tree. Whirls of snow cruised across the field in mini tornadoes, obliterating the boundary of the meadow and all sense of perspective. The land lifted and suspended us.

Someone yelled at us from the road, breaking our reverie, and Peter and I ran back to the car, making a hasty apology to an irate elderly woman dressed in three coats, a wool hat, and a clear plastic visor and accompanied by a honey-colored dog. We left in haste like teenagers caught making love in a field. It was terrifying—not the woman or the dog, but the way the land had enticed a kind of free fall into air.

Less than a year later, the hillside farm became our home. It wasn't for sale, exactly. And we weren't married or even certain we could make it work, exactly. It was all a giant leap of faith, a leap prompted by the energy of that wintry hillside that had uplifted and shaken us. The farm had been given to a land trust by a woman named Ann Day who had lived there for nearly fifty years—the same woman who caught us trespassing—and we competed for the chance to buy it from the land trust in a months-long courtship. The right stewards to take it over, they said, would be people who kept it in agriculture, preserved the old barns and farmhouse, and ran educational programs rooted in the land. Our proposal said that we would run a learning center for social justice and land justice leaders to come together around the work of restoring land and repairing relationships, and that we would grow organic produce, have a pick-your-own orchard, and raise sheep.

All through the application process, my nerves were on fire, my fear of the monumental responsibility we were taking on existing hand in hand with my desire. I wasn't ready, yet, to fall in love with a new farm. We would go to visit Ann Day in the old farmhouse that she ran as an inn, sleeping in a room with low windows under the eaves—furnished with a horsehair mattress, peeling yellowed wallpaper, and magazines from the 1950s—and I'd have disturbing dreams. In the morning, in a breakfast room filled with plants and piles of papers and warm eastern light, we'd hear about all the people and programs Ann had hosted over the years—community nature walks and poetry slams, night skiing and Spanish classes. More importantly, Ann had sheltered refugees from Central America during years of civil war, people with mental illness who needed support, and generations of families as inn guests who loved the place like home. We were stepping into a world—every part of which was well-loved and well-worn—that we were boldly promising to extend and honor for decades to come, with little idea of what it would require of us.

PETER AND I MOVED IN TOGETHER AND REMADE OUR LIVES around all the things we believed in and the promises we wanted to keep on this land. Peter—when he wasn't building tent platforms and outdoor showers or starting a nonprofit to run our programs—was on the road a lot, raising money, teaching and facilitating, giving talks accompanied by his evocative photographs about people's relationship to land. He has a gift for translating values into action, for inspiring people, and for creating programs and structures that shift movements for change. Building our livelihood from scratch around nothing but the conviction of our ideals, in a place that required so much labor and devotion, so much vulnerability and humility, scared me as much as it inspired me. It tested my introverted soul. I found solace in the pragmatic, in what I knew how to do with my hands. I planted vegetable gardens and made compost piles, pulled down miles of barbed wire, brought home sheep and chickens, and planted seven hundred blueberry bushes. For a while, I kept up my life in books, editing other anthologies and manuscripts as a freelancer. I moved my friend Brian's huge 1917 printing press from Putney School into the small hay barn so I could print broadsides. My first project was an invitation to Willow's fourth birthday party; she helped me find each letter and tie the block of type with string. Standing on a chair to reach, she cranked the huge wheel and watched the roller press out her name into cotton rag paper. The press didn't stay long—Brian had to sell it to pay his bills—but it remains part of the vision Peter and I share in which food, ecology, and activism are all intertwined with our love of art and words.

For the first few years, the promises we made and the immense responsibility we felt to the land visited us daily in the form of Ann, who had chosen us as the new stewards and was now our neighbor. Each morning Ann would walk down the hill to feed our chickens scraps of salad and fruit and to bring Willow bags of sweets. It was years after we bought the place that she and her children moved the last of their belongings out of the barn—old iron saddle racks, horse harnesses, salvaged lumber, collections of china, boxes and boxes of scrapbooks and letters. Although the place had become too much for Ann to care for, her attachment was marrow-deep.

One spring morning, a few months after Peter and I moved to Knoll Farm, we looked out our bedroom window to see Ann walking across the lawn. She wore a raincoat and baseball hat. She moved briskly despite her years, always with a businesslike authority. When she was younger she'd been a fearless and expert skier and sledder, and she had told us stories of swan-diving off high beams into the hay mow and riding bicycles into the pond for fun. This morning, we watched this woman we admired and were so grateful for stop abruptly, her back to us. We followed her gaze out over the sunken pond and its cataract of spring ice. We watched as she stood still for a long time, turned, and walked heavily away. The day before, taking advantage of the sorry state of the pond, which had nearly drained when the overflow pipe was split by a block of ice, Peter and our friend Grant (who was really a friend of Ann's and still lived, with his wife and Ann's cat, in an apartment in the small hay barn) had torn out the old red dock with its rotting pillars, replacing it with a large, single stone from the woods. It looked better, we thought, to say nothing of safer. The next day Ann told us that her daughter had been married on the dock, in the summer—there were damselflies, she remembered, and hundreds of barn swallows sipping insects from the surface. Her nostalgia was loaded with expectation, even accusation: Why did we have to change what she had always known as home?

We knew more of this unspoken story: It was a month before her daughter's wedding that her son had found her husband, Frank, in a deep ravine in the forest, where he had taken his own life. The two days of wondering where he was, the body in the woods, the wedding without a father, the dock poised over a pond, the morning dew on the cattails, the sound of the bell by the door, the dip of a swallow on the water, a damselfly's iridescent blue, the hellos and the goodbyes—all these were part of Ann's intimacy with this place, and not ours.

Peter and I both thought we saw Frank, about a week after we moved in, coming out of the barn door. Perhaps, as Lauren Groff wrote, a ghost is a person and a place dreaming at the same time. But ghosts have nothing over memory. Ann walked through these woods ahead of us, catching invisible spider threads on her face, in her hair. We, who were beginning to walk after her, encountered nothing but the clear air of morning, a path through the woods, the cadence of leaves long layered to make this very soil. Soil we had plans for—whose fertility we had yet to uncover, whose stories to us were still like seeds and not ashes. But, with all this, I understood Ann deeply, intuitively. She was passing her torch to me, generously and fully, but I knew how much it ached. I had to learn to carry it. I had to learn to steward this place well.

The sheep, it turns out, were what I needed to help me immerse in a new place. Along with the land itself, they have been my touchstone and my teachers. Sheep have helped me become a good shepherd, not just to them, but to a place that is my sustenance and joy as well as my unending labor and worry.

The idea of belonging—what it is, what enables it to happen or not happen, where and how it is found—has preoccupied me most of my life. My practice of being a shepherd is a huge part of my ongoing inquiry into belonging, not just my own, but understanding what it means for all of us in this time of profound loss of the natural world. This deep encounter with land, this process of falling in love with a place, was something I knew in my bones, something that seemed fated for me. And at Knoll Farm, I arrived with my begging bowl of perception empty, with room for what I experienced to nourish me in new and surprising ways.

PETER AND I WERE MARRIED IN 2003—A YEAR AND A HALF after moving to Knoll Farm, a year after getting sheep—in a ceremony that was as much about the land as each other. "We invite you to join us in celebration of our love and marriage to one another and to the land," read our wedding invitation. Willow, who was five, drew a picture of the three of us with flowers, holding hands, a huge sun coming up behind. Long grass. Below her image I printed lines from "Like Three Fair Branches from One Root Deriv'd" by Robert Hass: "We are passing through the gate / with everything we love. We go / as flesh, as fire, as marble."

At dawn on the summer solstice I walked from the farmhouse where I had spent the night alone to the top of the high pasture where Peter, dear little Willow, and our families waited for me by a ledge of bedrock. They had all camped that night in the dewy grass. Most were dressed in pajamas. My father, elegant as always in a pale pink robe with silk lapels that he borrowed from his partner, Nancy, walked the last part of the long hill with me to meet Peter and Willow. The moon was setting as the sun rose. Willow wore a white cotton sundress, and Peter had braided her hair. All through the ceremony she stood under and between us with perfect poise and great seriousness, looking up, our bodies like two swaying trees. Wren was there, too, an apostrophe inside me.

After the ceremony, most people went back to bed, followed by brunch on the lawn at long tables with hay bales for seats. Our friend Michael dug a pit where he slow-cooked a whole lamb barbacoa-style all day. He rubbed the meat with ancho chilies and salt and heaps of spices, then cooked it on the coals at the bottom of the pit, covered with agave leaves and a lid of sheet metal and buried. For our wedding feast, Michael flaked the lamb off the bone and served it with mole and fresh handmade tortillas, rice and salsa verde, roasted vegetables and salad. My grandmother, who was ninety then, sent a traditional wedding cake she made, all the way from England. We set up tables in the hay loft of the barn for dinner, with the swallows flying in and out, sun and sudden shafts of rain slanting across the huge open doors and bringing the smell of earth and sheep.

WREN, BORN IN THE FARMHOUSE DURING A BLIZZARD EIGHT months after our wedding and raised outdoors, is a child of this land. Willow, who went between her mother's home and ours each week, has always known what it was to move from place to place. Fully grown now, and more than the rest of us a creature of the modern world, she has made her home in the city. But in Wren I see the longing to let loose and yet also to hold on to what has made her who she is. Her umbilicus is wrapped around the roots of this hillside—an anchor in the storm, a tether around the ankle. For each of us, no matter what it entails, this dance of leaving and staying, known and unknown, love and loss, requires our whole heart.

Along the path, something unexpected often appears to guide us; the length of a drenched hillside and its humble flock of sheep continue to show me the way. As Giono asks in *The Serpent of Stars*, for attempting to pry open a glimpse of understanding into something so elemental and ultimately unknowable, *May I be forgiven.*

CHAPTER 3

THE FLOCK IS THE LAND

All the echoes from the hills trembled with bleating.

—JEAN GIONO, *THE SERPENT OF STARS*

It's early April, and the frozen heart of the land is slow to let go. The frost rises and cracks the ground, letting in rivulets of rain to penetrate and loosen the cold, and in the morning my boots crunch across infinite worlds of ice crystals and miniature prisms. On the south-facing fields where the snow has melted, the first shoots soften the hard dead planes of the hill with the slightest mist of gray green. The sheep paddocks outside the sheds where the ewes birth their lambs stink of manure steeped in rain, of rotting hay and bare earth beginning to waken.

Two robins who have just returned from the south peck at patches of mud between giant blocks of snow and ice that have fallen from the barn roof, white wing patches flashing with each hop. These are the early scouts; soon the fields will be full of robins, and we'll hear their beautiful full-throated music from the tops of the maples. This morning is quiet and cold, without birdsong or the chirr of insects or a tentative call of the first frog, but I sense that the long sensory deprivation chamber of winter in the north is about to break.

Around here, the sheep are the first to mark the season of voices rising. Grass cannot grow quickly enough for a ewe flock in spring. My sheep sniff the metallic blood smell of earth thawing and grass growing and baa loudly in their hunger. The ewes who have birthed are already rangy, transformed from wide-beamed and expecting to leggy creatures whose lambs suck milk from their udders and flesh from their bones. They eat a few mouthfuls of last summer's now-dingy hay and pace the perimeter of the winter paddock with their lambs close behind, crying at the fence lines as they beg me to let them wander and kick up their heels for the first blades of green. And I, having learned a shepherd's mind of pacing for the season, cruelly refuse to open the gates. I stuff some more brittle hay in the feeders, and they look at me with defiance, getting down on their knees to snuffle tiny morsels of the neon-green stuff through woven fence wires.

On a small scale, we are "grass farmers." We farm grass, essentially, and the more grass we grow, the more meat, milk, sweaters, and rugs we produce. To raise animals purely on grass—without feeding grain, say—isn't as easy as it sounds, especially in places that are arid like the Western Range, or below freezing for seven months of the year, like Vermont. For us, it requires cutting and drying enough grass to endure a very long winter and then keeping up with the full-on sprint of the grazing season by moving our animals almost daily to capture all we can of the hillside's grass. The sheep not only eat to nourish themselves, they magically create the next round's moveable feast as they go. Their hooves subtly and gently till the ground, disperse grasses, and stimulate dormant clover and vetch to sprout. They make pockets for rain to collect in dry ground and open the thatch in cold ground, and they leave perfectly pelleted compost in the way of dung everywhere they go.

All winter, I long for the grazing season to begin. It feels like freedom to wander with the animals after so many months of lifting hay bales and breaking ice from buckets while muffled in five layers of clothes. There is nothing like the excitement of feeling spring return. But I know from past mistakes that if I move them onto spring pasture too soon, the sheep will go from being instruments of fertility to a horde of locusts. If they eat the grass too low when it is just beginning to waken, the pasture will lose its strength and grow back too slowly when I need another round. The winter paddocks are a good example of this: nibbled daily in spring while the ewes are waiting for greener pastures, these small areas never grow good grass. It has been permanently stunted. So, my sheep, they wait.

THE RAIN HAS TURNED TO SNOW IN THE NIGHT. CLUMPS OF slush swirl in the meltwater of the barnyard swale. A channel the color of strong tea runs along the paddock, under the fence line, and down the long sloping fields to spread out near a weeping willow that glows yellow under its cloak of new snow. The willow tree and a nearby clump of birches act as wind- and water-breaks. Around their spreading limbs will be the first good grass to graze, perhaps a week from now, as the heat of the sun helps all that trapped moisture manifest and morph into food. Below the willow is an area of coarse sedgelike grass that grows in wet soil and has to be grazed early before its sharp, waist-high stems are unpalatable. Later, when it gets ahead of me again, I'll turn the flock in just to trample it and open some sunlight into that damp soil. Beyond the wet swale is a high knoll that has the opposite problem: Its soils are so thin that the sun keeps it dry and stony, and mostly low plants that can conserve their moisture by hugging the ground—cinquefoil and bluestem and strawberry—grow there. Over time, as the sheep build soil fertility with their droppings, I hope I'll see more timothy and orchard grass, dandelion and clover.

We are well into lambing now. The birthing began in late March, reached a peak about two weeks after the first lambs, and is now tapering off with a few stragglers in late April. A set of newborn twins shiver a little from cold in a corner of the shed, their mother licking a placenta that quivers jellylike in the hay. This instinct to eat the birth sac and placenta is probably a holdover from being a prey animal in the wild and trying to remain in hiding as a mother giving birth, but these membranes are also dense in nutrients and provide energy for the exhausted mother. I set up a couple of panels to keep them penned in the corner to make sure that the mother, a yearling, is bonding and has plenty of milk. I clip a heat lamp on a nail in the wall and set the lambs beneath its orange glow. Both are white, both female. I rub them with a dry towel, making a gentle pass over their small pink nostrils to make sure nothing is caked there. Their fleeces are stiff from the dried birth fluids that coat them like a film of egg white.

I dip their umbilical cords in iodine, holding the neck of the bottle over the belly button and swishing the dark-red liquid, then I take a pinch of delicate skin on the flank to give each a shot of tetanus antitoxin. I feel under the mother for her warm udder taut with milk and give a strong pull on each of her two teats to clear the waxy plug. The first milk, thick and yellow, splashes onto my boots. I hold her head still, chin tilted up, and squirt a dose of worm medicine into her mouth. Parasites can overwinter in the mother's gut and do no harm, but these internal stowaways are clever, and the mother's hormones during birthing stimulate the parasites to shed their eggs through the feces, so I need to kill them first.

I head outside to check on everyone else. On a clipboard on the wall is my list of each ewe who has lambed, the date, the number of lambs, and their gender, coloring, and sire. This way I can track the genetics and sell the offspring as purebred, registered Icelandic sheep. By selecting the genetic traits I want and pairing my best animals, I can sell lambs for a premium, which supplements our other income from meat and wool and is how we have started to make a decent income on a way of farming that would otherwise be marginal. Writing everything down helps me memorize each family group so that I know if anyone is missing. A missing lamb might just be running around the paddock, but it might also have its leg stuck in a wooden pallet tied over a hole in the fence, or have climbed the wooden lid over the water tank and fallen through a crack that was left open too wide, or have been snatched by a coyote. Or it may have been rejected by its mother and be trying to nurse from another ewe and not getting enough milk. Five years in now, all of these things have happened.

    When they see me, the ewes come into the three-sided shed to eat, shaking off heavy cloaks of wet snow like dogs after a swim. That's when I notice Leda. Ten days after birthing triplets this magnificent white ewe has transformed into a ribby, mangy she-wolf with haunted eyes. Something isn't right. I can't figure it out, because she is eating all the time. She is thinner than all the others by far, and her lambs bunch around her legs, butting and tugging on her udder in violent desperate bursts. She steps aside in pain. She has no milk for them.

    The three lambs, one black and two white, try to nurse from other ewes nearby but are butted or kicked away after no more than a taste. This fierce proprietary instinct of the mother Icelandic sheep to feed only her own young with the milk they need to live has survived thousands of years of domestication. I've seen in more modern breeds—in flocks where for generations the lambs have been removed from the mothers so that the farmer can capture the milk or fatten the lambs for slaughter—that the mothering instinct is diminished. The mothers don't seem to be bothered as much by a random lamb snatching a snack from her udder. But the Icelandics and many other primitive breeds have kept the instinct to feed and protect only their own, giving their offspring a better chance of survival. Only once in my own flock have I seen two ewes share the care and feeding of their collective four lambs. These two ewes were themselves mother and daughter, and they gave birth at the same time, next to each other in the same lambing shed. The newborn lambs as well as the two mothers lost the trace of origins, which are imprinted through each mother's and lamb's unique signature of smell—birth fluids and saliva and milk—as well as the sound of both voices.

I decide to put Leda and her triplets in a stall inside the barn where I can watch her closely. Our barn was built as a cow dairy a hundred years ago, and nothing about it is convenient for raising sheep. We have cobbled things together in classic Yankee "make-do" fashion. The shed where Leda is eating hay, and where I delivered Blue's lambs one night a couple of years ago, was originally a holding shed where the cows' manure was shoveled out of the dairy. It is three steps down from the main barn. Sheep will overcome their fear of climbing stairs, I've learned, only if confronted with the greater fear of losing their lambs.

Moving lambs is a job Wren, at age five, loves to help with. Kindergarten is three days a week, so she's around the farm a lot. I try hard to set aside time when we can simply play or do what she most wants to do, but I confess that a lot more of my energy goes into trying to fold her into my daily (endless) chores. I like to have her around.

She has learned to entertain herself with absolutely anything that's at hand—a stick, an ant, a toad—but inevitably she can get tired and cranky, especially if it's taking me hours to do one task, or the weather is foul. It is a small triumph when something exciting for her matches up with something I need to get done.

As I'm learning how to be a shepherd, and a mother, I draw so much, without even realizing it, on what my mother passed down to me: her patience with animals and with me (she had a weird amount of patience for everything, my mother), her care in explaining what she was doing and finding creative ways for me to be part of it until I was old enough and encouraged to do it on my own—to have my own calf or my own patch of carrots.

"He's a little hefty," I say, handing Wren the black lamb. "Hug him close."

I hold the two white ewe lambs clamped together with one hand under each warm belly and dangle their loose and leggy forms at Leda's nose level so she will follow. Sheep have bad eyesight, and the minute she can't smell or hear the lambs, she will get wild, charging off to find them.

"Maaaa," Wren says. "Over here, mama mama."

We walk slowly, backward, up the steps. Leda hesitates before jumping up to follow. We dangle the lambs to lure her along the dark aisle, over the heaved and broken floor, past wooden cattle stanchions covered with cobwebs and piles of buckets and bags of wool. Her fear is palpable, but so is her desperation to follow the smell of her lambs ahead of her; each time they struggle and let out a little mew she rushes forward, until finally we get her into a lambing pen on the far side.

FOR A COUPLE OF DAYS I FEED LEDA EXTRA HAY ALONG WITH some alfalfa pellets and beet pulp soaked in warm water, and I give her salt and minerals. I make a molasses and milk thistle tea and massage her udder to stimulate milk letdown. Nothing helps. I watch her eat everything voraciously and still jettison weight like someone on *Survivor*. Since she has nothing for them, we supplement the lambs' diet with formula from a bottle several times a day and watch their middles swell like tiny potbellied pigs, but they are no longer growing in visible leaps like the lambs in the paddock with their mothers.

I'm running out of ideas when one morning I see Leda droop her head and spit out a perfectly round disk of green cud, a gob the size of a silver dollar. It gleams in the hay, rejected, back where it began, and yet it is distinctly bright green, as if Leda's digestion has restored her dehydrated meal to a seaweed salad. For all the hours I've witnessed the sheep resting and chewing their cuds in rhythmic circles—watching the lump of cud make its way down the throat with a deep swallow, then come up again with a subtle burp—I have never actually seen the cud itself. The cud is like a coin in a vending machine: It makes the machinery go and results in food that can be digested.

To chew one's cud, or to ruminate, meaning to mull things over, is an expression that springs from a basic digestive process that sheep, cows, goats, and other cloven-footed animals with a rumen (therefore: ruminants) simply cannot do without. The brilliance of these animals, and undoubtedly the main reason humans domesticated them, is that they have a vast and complex cadre of microscopic flora in their gut to help them break down the tough cell walls of plant material and turn it into protein and nearly every other nutrient they need to grow and to make milk for their young.

Rumination begins with chewing. Lots and lots of chewing, and mixing the grass with the fine-tuned tincture of the saliva, which contains digestive enzymes. This green slurry then makes its way into the rumen, a miraculous fermentation vat where anaerobic bacteria, fungi, and protozoa harvest nutrients and provide them to the animal's bloodstream in usable form. A healthy sheep is not so much an individual animal as one member of a complex, highly evolved biotic community, much as we now understand a tree to be one element in a vast interconnected and communicating forest ecosystem.

The same is true for us. We also rely on a vast interconnected world of invisible life to do the work of digestion for us. Though our gut flora are not adapted to break down bark and hay, they are no less important. What I have learned studying my sheep is that herbivores eat plants to feed the microbiome in their gut, and those microbes in turn feed them. These microbial friends aren't just an army of recyclers breaking down packets of food so they can be digested. They are a large part of the vital food themselves. As microbes—who are short-lived—die in the gut, they provide up to 80 percent of the protein needs of ruminants! In essence, no one is vegetarian; invisible bodies outside and inside us are constantly doing the ordinary, utterly transformative work of dying to nourish life.

Don Mitchell writes, in one of my absolute favorite books, a fable about sheep called *The Souls of Lambs*, "The peaceful ruminating sheep is not digesting grass; rather she chews to help flora digest her grass for her. She is the white-robed brewer stirring up a frothy wort." Leda is spitting out her frothy wort before she can chew it enough for her rumen to accept it. She is missing not only the nourishment of her cud but also that peaceful, rhythmic ritual of digestion that sheep do throughout the day and night—and that has everything to do with the stillness and sense of ease required to let down her milk and feed her young. It's all a great circle of well-being that, when interrupted in any one place, no longer turns around.

Observing that sheep have to chew nearly every waking hour of their lives, either eating or re-eating their cud, I wonder if her problem is that the act of chewing itself is giving her great pain. I call Roy Hadden, whose vet practice is just down the road. Roy sees pets and does surgeries in his clinic every morning, then in the afternoons he does rounds to farms, though less and less. When he started his practice about twenty years ago he took care of nearly thirty dairies, he told me. Now he has three.

Roy strides in, hoodie hugging his muscular frame, baseball hat crammed over his buzz cut, his eyes already looking past me for his patient. I've almost stopped being intimidated by him, for his brusqueness and confidence overlay a deep kindness and a love of mentorship. Each time he comes I learn more, and he's begun to trust my novice opinion about my animals' health, leaving medicines out for me at the clinic for basic problems like mastitis or pneumonia, without having to visit.

"So this is our patient, and she's not eating, eh?"

"She's eating, just spitting out her cud."

"Sure looks like she's not getting much."

Roy pulls out a stethoscope, starts palpating Leda's abdomen, then takes her temperature. The lambs are crawling up his legs and sucking on his clothes.

"Hungry little buggers, aren't we!" He grins. "Now you . . . what do you think you're doing . . . you're being all kinds of help," he says as he gently lifts a lamb off his arm.

He has me pry Leda's mouth open as wide as I can while he shines a flashlight inside. She thrashes her head, and I have to pin her against the boards with my whole body to hold on.

"Yup! Busted molar in the back. Razor-sharp too. That's making it too damn painful for her to chew."

Roy does his best to file the sharp edge of the tooth while I keep holding her mouth open. Her warm saliva runs under my cuff and down my arm.

As Roy packs up to leave, he tells me it's up to me. The filing might help, she might get by, or we could schedule a tooth surgery, but I would somehow have to bring her to the clinic for him to do that properly. And it would cost a few hundred dollars.

We walk out to the truck. "Thank you. I'll let you know," I say.

A EWE'S MICROBIOME DEEP IN HER FOUR-CHAMBERED STOMACH conditions grass to nourish her own flesh. But that's not all it does. Like the sourdough starter bubbling on the windowsill or the thick mat of scoby in the vat of blueberry vinegar I make each year—both of which we call the "mother"—a ewe's inner microbiome is also the digestive starter kit for her young. Her lambs are not born with intact rumens or the ability to digest grass. Their digestive flora is created in part by their mother's first milk, called colostrum—a thick yellow, highly medicinal, antimicrobial, immune-building, and downright magical fluid—as well as through her saliva, skin, and feces as she licks, nurses, and lives with her young.

Taking young animals from their mothers—universally done in modern farming—predisposes them to sickness. Industrial farming practice is to feed antibiotics prophylactically, from birth, and it's hard to find a milk formula for young animals that isn't medicated. Keeping a lamb or calf with its mother even for the first four days of its life, when the mother's colostrum is flowing, has a big impact on its future health and growth. Keep mother and baby together longer, letting them graze together, and the mother can be the link to generations of learned and embodied knowledge about that landscape, what to eat and when, where to walk to connect fields, and even where to find the safest place to lie at night.

Fred Provenza, a scientist-sage when it comes to the behavior of grazing animals, writes about how this knowledge of landscapes, passed down from one generation to the next, is essential for offspring in pastoral systems to survive. The mother's influence begins in the womb, "through flavors of foods she eats in her amniotic fluid," and continues after birth as she shows her lambs what to eat and not to eat.

Industrial farming, in which animals are fed grain in barren feedlots, interrupts millions of years of coevolution—for the grasslands and grazing animals that formed a unique relationship, for the humans who understood how those relationships ensured mutual flourishing and could turn grass into milk and meat for their own consumption, and for an entire universe of soil and gut microorganisms that are designed to help soil grow grass and ruminants to digest it.

As I learn how everything is connected, I see more clearly how thin the line is between damage and healing in farming. The way we farm places us on a spectrum from poisoning and depleting the land to being a partner in its restoration and resilience. The difference seems to have a lot to do with observation of natural systems and a respect for the land, seeing it not just as a hunk of rock and dirt with three elements (N-P-K) and plants and animals growing on it, but as a symphony of interconnected parts that have collaborated and improvised over millennia. When it comes to land, the whole is, in fact, infinitely greater than the sum of its parts.

THE WEATHER HAS TURNED COLDER, AND THE SNOW IS STICK-ing around. The hills recede behind the falling clouds. Soon I will feed the last bales of hay from my winter supply, like a savings account carefully doled out to last. But this year spring is late to arrive; I don't expect to be grazing before May. I call Elwin, my neighbor down the road who has a small beef herd and makes beautiful hay and the best sweet corn in the valley. He is close enough to drive over with a round bale clamped in the mandible of his John Deere and place it in the paddock. The round bales he makes and wraps for the winter are semifermented by spring—pungent, leafy, full of protein, and smelling like malt. They make the ewes wild.

Elwin is tall and lean with a handlebar mustache and sun-burnished eyes that are always smiling. He grew up on a dairy farm in our valley and, when his father no longer wanted to carry on, took it over and turned to raising beef. The economy for dairy had been on a long downhill slide by then. He grows all his own feed corn and hay on beautiful fields along the Mad River and sells his pasture-raised beef and pork locally, to a ski area and a couple of markets. You can stop by his farm stand, too, a cramped wooden shed full of old papers stacked on a metal desk, two freezers full of meat, a fridge for duck eggs, and an old milk churn with a slot in the top for stuffing in your bills.

"How's the battle going?" I ask Elwin as he climbs down from his huge John Deere. Despite the cold wind he wears only a worn flannel over a T-shirt, no hat, jeans, and steel-toed boots.

It's the first time I've seen Elwin since a freak cyclone tore through his cow barn last fall, leaving an enormous ragged hole through the roof and walls, as if a giant mouth had chomped through sheet metal and beams and one hundred years of rusted machinery and dust to cleave the barn in two. After blasting a path through the barn, the winds hit the forest beyond and knocked down dozens of pines as if they were bowling pins. You can see the hole in the forest from the town road. Nothing else in town was touched. It was like God had pointed a finger at him.

Elwin has spent the spring hauling all the trappings of his lifetime and his parents' lifetime into a giant burn pile, torching it day after day.

I say how sorry I am about him losing his barn, but he doesn't answer, and I sense his masculine pride. There is no room for pity.

"It's going," he finally answers. "Nice to be up here. Where would you like this bale today?"

He's twiddling the ends of his mustache and looking amused, though I can't think why exactly. He must have a trillion things to do, but he always takes time to stop and chew the fat, as they say. I deeply appreciate Elwin and several other farmers who are half a generation older than me and still going strong—Hal, who brings us hay from Lincoln and makes fifteen thousand bales each summer on his own; Easty, who makes maple syrup and, like a mountain goat on his tractor, has been fearlessly clipping the steepest slope of bracken here for something like fifty years; Doug, who owns the local feed store and is always a source of help with equipment. I appreciate their work ethic, their knowledge and kindness, and most of all their stories, which are often about wildlife or weather or equipment failures. Disaster stories—getting pinned by a steer, bears destroying the corn, a trailer shearing off the back of the truck on a steep road—but also peaceful stories of stopping the tractor to watch foxes playing with their kits, or finding glory-of-the-snow bulbs around an old cellar hole. I often say that when their generation of farmers retire, I'll give it up too.

Once, Elwin came to the farm with a limp. I hadn't seen him for a while. I asked him if he had hurt himself, and he told me his leg was bolted together in six places, or something like that. Everything about the story was extreme: Skidding maple logs out of the woods that winter, one got hung up in the forest, and when he cut it loose it whiplashed him to the ground, the trunk pinning him across the legs. His chainsaw was still running and still in his hand. His flip phone was still in the breast pocket of his flannel shirt. He cut himself loose with the chainsaw, called his son, then lay there on the ground in the snow until the rescue crew got out to him because his arms were the only thing he could move. He had a punctured lung, a shattered femur, and a bunch of other injuries that would, in themselves, make a story for most of us but were insignificant in the face of such disaster. He told the story like a badge of honor. By the time he was only half healed, he had gone back to doing nearly everything on his farm by himself, working sixteen-hour days. I'd never heard him complain about anything except the weather.

As Elwin unloads the round bale in the paddock, we can barely hear ourselves speak for the crying and carrying on. The sheep swarm the pungent bale and stuff their faces into the fermented grog of summer's past. A lithe badger-face yearling ewe leaps to the top, where she can gorge alone above the others' heads.

The lambs pick up the excitement and fly around the paddock in a fluid gush like a murmuration—the big kids competing out in front and the youngest veering off sideways as if not quite understanding the game. The lambs are getting more independent by the day, like toddlers venturing from mom's side to poke their heads and limbs into curious corners or play running games, while the mothers turn to the business of eating and conversing with other mothers, only to realize now and then that their lambs have strayed. Several ewes pull away from the fragrant hay bale to call loudly across the paddock, shouting with alarm at the babes to come back. The lambs, who have raced to the top of the huge windrow of compost and are jousting for position as king of the mountain, ignore the calls. Bluestem, in the role of playground monitor, charges up the pile herself to lay down the law.

Later, as the day wanes and bellies fill, the flock will fall quiet. Now and then a lamb that is lost and hungry will call out, looking for a nipple or the chance to lie down next to a warm familiar body as it grows dark. The word *plaintive*—from the French *plainte*, a lamentation—is the only word that comes to mind for that small voice of the lamb. A *lamb-mentation*.

ALL THESE LAMENTING LAMBS REMIND ME OF A TRIP TO Majorca with Peter right after we were married. We took Willow, who was five. Our wedding present from a friend was to stay in her father's home, a rambling stone farmhouse that dated back to the 1400s and seemed to be part of the mountain itself. One ducked out of the dusty sunlight though an arch, down stone steps cupped by use, into cool corridors, enveloped by the smell of lavender and garlic, olive wood and dry dung. Through the house, in the back room—if that's what you could call it—was a cavernous stone gallery that still contained a centuries-old working olive press, complete with a granite grindstone twelve feet across that was rotated by a horse hitched up to a pole. The olives were ground, pressed by the massive stone, and the oil ran down a long wooden chute, gleaming and green, with a fragrance more complex and mysterious and alluring than any I'd ever experienced.

The house was surrounded by parched hills, the farm's ancient olive groves, and sheep. Lots and lots of sheep. They roamed about the farm and way up into the hills without fences, wearing bells that rang all day and night with their coming and going. It was June when we were there, and the lambs were half-high to their mothers. The spring rains had come and gone, and I don't remember seeing any grass. Brown leaves lay curled under the olive trees and crunched under our feet as we walked above the villa to a rocky mountain traverse. Even there, sheep trod hard the paths in single file through the pines. I couldn't fathom what the sheep could find to eat. At night, too, we heard their crying and their bells jangling, and I would lie awake where we slept on a bare mattress on the roof, listening to the olive trees rustling in the dry and heavy darkness, fretting for them. I'd roll over and see that Peter's eyes were open and reach for him. Little Willow was the only one who seemed oblivious to the lamb-mentations and slept, glistening with sweat in only her underwear, her hair stuck to her cheek.

Chances are good that my fretting came from my lack of literacy in the language of the land and flock in that landscape. These sheep, though vocal, did not look malnourished; I supposed they were starving based on my assumption that only a green landscape possessed food. Perhaps my sheep would have starved there, but these isolated Majorcan sheep and the intelligent microcreatures inside their guts had evolved to make the most of eating drought-tolerant, extremely nutrient-dense shrubs, high in tannins and fiber. Now, with modern sheep breeding designed to raise animals for uniformity and for one purpose (meat or wool), many ancient sheep breeds adapted to their terrain around the world are disappearing. Once, all sheep flocks reflected the habitat in which they evolved, the plants they ate, and the climate they lived in, sometimes so precisely that local people could confirm when they had crossed into a new mountain range or region because the flocks of sheep grazing the fields would look distinctly different. Just as the animals evolved with the diet presented to them, the plants evolved back, and it's possible that no other sheep flock and no other grass species would do as well there. The natural world, which includes the farming world, is in a constant dance of cocreation that we disrupt so easily because we have been trained to think of it as separate and interchangeable parts, as in a machine.

For a few days, I continue to watch Leda and feed her the equivalent of a green smoothie, but without the ability to chew her cud, she is doomed. The three insatiable bottle babies find new ways to get out of the pen every day to follow us around the gardens and the orchards as we try to work, sucking on our sleeves and pawing us with their sharp little feet. Something has to be done.

I saw a documentary years ago about a shepherd in Morocco whose camel is sick and won't produce milk for its calf. The family calls in a shaman to heal the camel, and the shaman chants, waves incense, and rubs things on the camel as it lies in the dirt, doe-eyed and listless. They sing and sing for days. Finally the camel dies.

It was a boring film, and most of the audience left during the first day of singing, never getting to the punch line. Well, there wasn't really a punch line, but I was deeply moved to see the lengths to which this family went to save their animal. They knew that this animal's life was essential to its offspring's life, which was essential to theirs. The film wasn't worth watching, perhaps the American audience decided, because no farm animal was worth this trouble, and besides, nothing was happening. For me, a lot was happening in that film. All around the family's sunlit compound with their camels and incense pots was the thrum of commerce, new roads, trucks honking, deals going on. The shaman singing for the milk was also singing for an ancient lifeway and ways of knowing that were dying, being crowded out, dispossessed. I was left alone in the theater, feeling silly wiping off tears.

Deciding what to do about Leda becomes a decision between respect for the discipline of farming and reverence for her life. Leda isn't my pet, which blurs the lines for me. I know the cost of having her tooth surgically removed will wipe out all the profit of selling her lambs, and with farm animals, surgery doesn't always work, so I might lose her life as well and waste a precious $300. I am trying to farm to make a living, as well as to improve our land, but I also don't ever want money to make me heartless. It seems likely to me that much of human morality was discovered in our relationships to animals as well as to each other. And yet, in very recent history we have carried forward only our economic relationship to farm stock, and abandoned our moral one. I decide to try the shaman, whose name in my world is Roy.

The next day, I take the beautiful Leda and her three lambs down the road in the pickup truck and sit outside the vet clinic until the surgery is done, then Roy and I carry her limp, sedated body back to her desperate babies. Soon thereafter, Leda's lambs can smell her milk again and are wagging their tails under her belly, taking less and less of the bottles we offer them. Best of all, Leda lies down in the sun with the other mothers and peacefully chews her cud.

With industrial agriculture it's easy to stop seeing the connection between farming and the rest of the living world; in fact, the economic drive makes that an imperative. Wildlife, weeds, pests, fungi . . . all the elements of the living world are problems to be controlled. To go from damage to healing in our relationship to nature takes a place-based attentiveness, I'm learning, more than anything else. James Rebanks, in his beautiful book *The Shepherd's Life*, begins with sharing what it means to be "hefted" to a place, a word from his landscape of northern England that means for a flock to become familiar to its upland pasture, to belong there, to know and be known. The sheep in the pasture tastes and smells the unique array of plants there, feels the wind direction and strength, knows the smell of the soil and the rock and the way the shade wraps around the hill and moistens the grass, remembers where she rested unafraid. In this way she is hefted. Mothers pass this knowledge on to their young. All the species in this landscape have not only coevolved but are constantly adapting intelligently to change in order to find a way back to wholeness and equilibrium. My sheep and I are part of the circle and can do damage or do good. The only way I'll know is by noticing. The flock is the land, and the land is the flock.

CHAPTER 4

SPRING MEADOW

> The clots of sheep were on the move like clouds in the grass.
>
> —JEAN GIONO, *THE SERPENT OF STARS*

IT IS ONE OF THOSE PERFECT SPRING MORNINGS WHEN A thin mist hovering over the ground magnifies the bright green of the world, and small birds dart in and out of the soft veil with great purpose and lilting song. The new leaves on the sugar maples behind the house look like the crumpled trembling wings of a bat, poised above tasseled blossoms of the most delicate red and gold. Ferns are unfurling from the leaf litter on the edge of the lawn, papery copper skin splitting open to reveal tight green galaxies. The overflow from the farm spring rushes through a patch of red-stemmed pussy willows bursting with yellow pollen and pulses into the small pond between the house and the barn, splashing over stones and drenching the first bright whorls of field mint and pennyroyal. The pond exhales a fine mist, touching everything that is newly born, shiny, and ephemeral.

I was up before dawn to check only on lambs, but in this light, everything calls out to be noticed. I make a round through the blueberry field and apple orchards to the north of the barn. A male bluebird lands on the roof of the outdoor oven, rosy breast to the sun, and a house wren cocks her head at me from a nest in the blueberries.

Rhubarb pushes up bold folds of leaves from bare earth, and the flowering plum trees hum with a cloud of tiny black mining bees. The century-old apple trees, some of them no more than a hollow rind of bark with a few random limbs, are covered with downy gray-green leaves and a pink blush of blossoms.

When I finally come inside the house, it's late. From the kitchen I hear the creak of the school bus as it brakes at the end of our steep driveway, then passes on without stopping. Peter's boots are gone, and I know he's already up in the woods, cutting some trees that came down across the logging road. When I run upstairs to wake Wren for school she is already up, and she begs me to let her stay home. Her bedroom window overlooks the pond to the south, and the light on the water casts a shimmer across her ceiling. She knows it's splendid outside, not to be missed for anything.

"I'll watch the sheep all day and keep them from running away," she begs.

I pause before I answer. I am not that popular with her first-grade teacher at the moment, given the number of tardy notes we have accumulated when mornings got away from us in the chaos of lambing season. Peter and I never shy away from letting her skip school if there's a good reason, and she's already an expert at entertaining herself while we work around the farm, but when she's done with being independent, she's done, and I have to be done too. Having her home scatters my attention even more than it already is in the madness of spring on the farm. But she has an idea that captures my imagination too. A free-range day.

"Okay . . . let's try it," I reply after some thought. "Today I could use a serious and expert shepherd."

BIRTHING SEASON IS OVER NOW. FOR THE FIFTY-THREE LAMBS all marked on the clipboard in the barn, only two are marked "died at birth," which means it was my best season so far. It's hard not to fall in love with every one. Each lamb has its own coloring and personality. Some are white with golden faces or circles of black around their eyes, others are all black with a white dot on the forehead or tip of the nose. Some are badgers—dark underbelly and dark stripes under the eyes with silver or butterscotch on the blanket. Some are the opposite, called mouflon—light under the belly and in a stripe up to the chin, like a tuxedo and white bowtie. This is the most ancient pattern from the original ancestor of all sheep, passed down for thousands of years from the wild Asiatic mouflon. My favorite are the moorit lambs, ranging in color from dark chocolate to honey, and sometimes with lips and ear tips frosted with gray. "Sugar lips," we call this, and it means that the lamb will be dark brown when young but grow up to be light brown with a silver undercoat.

Today will be the flock's first full day on grass. I've been transitioning their diet slowly, with a few more hours of grazing each day over a week, so that their stomachs can adjust without dire consequences (imagine going from eating only toast for months to suddenly eating only fruit salad). At this point, I'm eager for them to live in the fields as soon as possible. Two of the lambs have dark diarrhea that could be from the new grass but might also be from coccidia, a microscopic parasite that lives in the soil of the winter paddocks where manure has built up. Lambs are susceptible to it and can die if not treated. I've sent a poop sample to Roy's office; if it's positive, the treatment is to dose each lamb with liquid medicine for seven days, which would require penning them with their moms and keeping them inside a week longer—all somewhat daunting right now in the general mayhem of the lambing paddock and counter to my impulse to get them out of there.

Thinking about coccidia and other invisible parasites is what currently keeps me awake at night. Lambs, especially, struggle to cope with the huge menu of internal parasites that they inevitably ingest with the grass. A whole cast of microscopic villains lurk in the sheep's daily salad. Stomach worms, tapeworms, roundworms, and the most deadly character—barber pole worms—all survive the digestion of the rumen and attach to the sheep's stomach lining, causing minute sores that, in enough quantity, lead to anemia and even death. Though the damage is slow, the death is sudden. Lambs seem fine one day, lethargic the next, then they're dead, with a bloody foam at their lips. Last summer I learned how to check their eyelids for anemia and dose them with wormer and B vitamins if they start going downhill.

Other parasites work more slowly, feeding on the animal's liver or lungs or heart. Liver flukes, lungworms, and a particularly terrifying creature, the meningeal worm, which infects the spinal cord and the brain, all reproduce inside the body of freshwater snails. Each morning I see scores of newborn snails hiding in the silver dew-soaked grass, and it reminds me how many trillions of deadly parasites I can't see, all in the iridescent beauty of the spring field where my lambs feed.

Today, I'm thinking as I start breakfast, *we will take the sheep somewhere they have never been before*. Land that hasn't seen sheep for decades has fewer parasites to infect them. It's also more likely to be overgrown with shrubs with medicinal properties for the flock's digestion. Thanks to Wren wanting to stay home and watch sheep, a vision for the day is forming: A long health walk for the sheep, a pasture renovation project, and an adventure for my eager shepherd-in-training. A walk uphill and to the east.

For decades, perhaps generations, the biggest hill above the farm was where Ann Day grazed her horses and Scottish Highland cattle. Most of the grasses died out from being continually grazed, and only plants with a bitter taste, plants with thorns, or plants that crept close to the ground survived. Parts of the field were like a mini dust bowl. Walk over that area now, and you'll see that goldenrod, milkweed, blackberry and raspberry, cinquefoil, bracken fern, and small saplings of poplar and maple are having their heyday. The hillside is too steep to mow, so only by grazing sheep or goats who are willing to eat these "weeds" will we help clear the way for light to reach the ground and germinate the more desirable pasture grasses and clovers that are still lying dormant there from years past, waiting for a chance. When the brambles and bitter plants like goldenrod are tender and young in spring, the sheep will eat them to the ground and begin to knock back their spread. At least that's the hope. Today we will experiment.

As I tell Wren the plan, she sits on the kitchen counter, cracking bright-yellow eggs onto a blue plate. I soak slices of bread, sprinkle in a little salt and nutmeg, swirl a slab of fat around the hot pan. She cranks open the window above the sink, and the delicious smell of cool damp earth mingles with the smell of browning butter. While Wren is with the sheep, and the sheep are renovating the pasture, my farm hand, Mary, and I will work on removing the greatest danger to having them free-range there in the future: barbed wire. Old barbed wire is everywhere at this old cow farm, and our sheep, who are ever-curious foragers and can't feel the first prick of wire through their fleece, have ways of finding the wire even when we didn't notice it was there, in a hedgerow or clump of trees. It binds them up like a sharp metal lasso. We've removed it from all the lower fields but not the upper acres. It's a good plan—nowhere on my long list of plans, but today Wren will be free, and so will the lambs.

I pack a bag with water, snacks, a wire cutter, and other fencing tools. We snarf down our French toast and head outside. I call the sheep together at the gate. The sheep come to know us by our clothes and our bearing, how we move, and they are beginning to know my voice. It's always the same call I make—a long warble from high in my throat that comes from somewhere alien, or maybe from the origins of time. I feel a little ridiculous when I call, but it's the one sound that makes them all pick up their heads and come running, from every corner of the field.

The sheep charge out of the paddock, so eager to taste the grass that they snatch bites as they run, forgetting for a moment about their small lambs who follow behind, crying in the melee. Wren has grabbed my tall shepherd's crook and is still dressed in her too-large, hand-me-down pajamas, dark blue with shooting stars and ringed planets. The wind is stiff, so she has my wool sweater pulled over the top. Her fine hair is growing long and hangs unevenly in her eyes and small, impish face. A plastic beaded choker she made on a piece of yarn lends a flair to the whole getup.

As we walk, I tell her a little about what to watch for when herding, although like a good farm pup, she already seems to know by instinct. "Watch their heads . . . they will turn next in the direction they are looking. Walk their flank, moving closer if you want to push them ahead and in a wide arc if you want them to turn. Sudden movements or coming straight at them might cause them to bolt, so we stay parallel. Stay calm." Our sheep are a primitive breed and not handled a lot, so they have a large flight zone. We don't have to get too close for them to react to our movements.

When Wren was very young, we didn't give her any tasks on the farm. We just tried to keep her from wandering too far away, from being chased by the mean rooster or clocked by a ewe who thought she should guard her lambs from a small mammal on two legs with a wide-legged gait and grasping hands. Almost as soon as Wren could walk, she would stand in the sheep shed with a sweaty fistful of hay, offering it to every animal within a few drunken paces. She did get pushed down a few times by a defensive ewe. And once, much more terrifying, the rooster flew at her with his spurs out, so Peter migrated the magnificent, evil bird way up into the woods to spar with the coyotes.

Wren learned to walk with Ruby, a moorit badgerface lamb born on a snowy night. Ruby was so cold when I found her that she was too weak to nurse. Her twin had already died, but Ruby lived in the house in a cardboard box close to the woodstove for a few days while she gained her strength, and once she did, Wren followed her unsteadily around, one hand on her back, both of them falling and getting up again on the slippery kitchen floor.

Willow began farming with us at five with a healthy skepticism about animals, which only grew when her own pet lamb, Boston, died unexpectedly of an infection, and her new chicks were viciously beheaded by an ermine, right through the tiny squares of their crate. It was terrible luck, and we couldn't often coax her to the barn after that. Today, she's at school where her mom lives, a couple of towns away.

As Wren and I walk east into the early morning sun, we feel the cool mist wetting our skin and hair. A robin throws back its head and sings from the very top of a wild apple tree. Our pant legs are drenched and heavy with dew. The sheep stream ahead of us, calling to each other in the yellow buttercups and delicate blossoms of lady's bedstraw as the early morning sunlight fractures the mist into white and blots out the terrain. We walk after them, as if through a light-filled doorway in a dream. Wren skips ahead, and I feel as if I could walk forever, right off the milky-blue cloud edge of the world.

    To the south of us is the steep slope down the fields to the forest, then to the wide valley where the Mad River winds north out of the Granville Gulf. Above the river and the layers of mist rise the eastern foothills, a tableland where the oldest farms, like ours, sit halfway up the mountains. The farms we look out on nestle below the Northfield Range, ancient mountains with a profile like a pregnant woman lying on her back—Mount Scrag her rounded belly, Mount Alice her breast, to the north a chin and forehead, to the south a knee. Rising like islands above the chop of fog are other mountain ridges to the south and west, while the forest and wind bring their song from the north. I see Donny Joslin's old sheep farm crumbling into the flinty hillside, and just down the road from there Gib and Sue Geiger's stately red barn and clutter of beeyards. Farther still, behind a foothill, is the Von Trapps' dairy, one of the few remaining in the valley, where they sell raw Jersey milk and make cheese. We are one of two farms on the west side of the valley. About a mile above us was the Vasseur dairy, sprawling across several hundred hilly acres of fields and a magnificent forest of sugar maples where they still have an active sugarhouse, although they sold the land in little lots after the patriarch of the family developed bad lungs from so many years in the cow barn.

Wren and I steer the sheep above a clump of white birches that grow out of one of the many rock middens scattered around the farm. Birch seeds, being so fine, are caught by the stones and germinate in bits of soil between them; in the winter when the birch catkins break apart, the millions of seeds look like the silhouettes of tiny swallows flying across the snow, dispersing into hollows to wait for spring. These large mounds of small stones are evidence of generations of farmers picking rocks behind the plow and tossing them aside. The fields are so steep and scattered with so many hazards in the form of protruding bedrock that it's hard to believe that previous generations of farmers chose to plow, but we know from an old farm journal from the late 1800s that Ann Day gave us that this farm produced mostly wheat and potatoes, and both crops require tilling. I imagine how walking behind a horse steering a heavy plow on these tilted acres would be mighty strenuous but somewhat less terrifying than driving a tractor across it.

A fox family has made a protected den in these rocks beneath the birch roots. Small bones are littered on the hard-packed soil around the hole. The sheep are grazing and wandering at a less frantic pace now, so we stop to look. We spot a mouse skull—picked clean—some other small bones and bits of gray fur, and a ribcage that looks like the recent remains of a barnyard hen. I knew this hen. The magnificent mother fox, her tail like a white-tipped matchstick coming through the grass, had stared me down the day before at dawn, looking to catch any chickens that had been roosting in the sheep shed rafters rather than safely in the coop. She had already found one and was back for more. I saw a trail of white feathers, dragged over the rails of the gate, resting upcurled in the grass like the lost pages of a letter. Foxes will cache-kill in spring, stockpiling as much game as they can in one hunt and coming back for it to feed their young over several days. Last spring, even with an electric fence, we lost twenty-five hens in one fox frenzy, in the middle of the day. Feathers were everywhere in piles, white and gold on the blood-soaked grass.

WREN HAS SHED HER SWEATER AND HEAPS IT ON ME TO CARRY. Lighter now, she clatters down from the loose stones and races up the hill where the sheep have gotten way ahead of us, nearly to the edge of a meadow where it meets the trees, fanning out into the emerging thickets of ground raspberry, blackberry, and goldenrod. *Will they eat this bitter prickly stuff?* I trail behind, trying to identify the plants beneath my boots. I recognize yarrow, whose Latin name—*Achillea millefolium*—refers to Achilles, who was said to use yarrow to stanch bleeding. I have chewed this plant to pack into a cut when I'm in the field. I also see plantain, another herb good for healing wounds. These are ancient plant medicines that every shepherd used to know. Would such herbs be good for parasites that cause internal sores? Will the lambs learn from their mothers what is good for them to eat, as I've read? These weeds in this neglected field may turn out to be my medicine chest, which makes me wonder how far I want to go with "field improvement" for lusher, more even stands of grass. Protecting some wilder, more overgrown places feels important—its own kind of stewardship.

At the top of the hill—so steep that in the 1950s it had a rope tow and was the valley's first ski area—is our "refuge village" of canvas tents and a cob bathhouse. The bathhouse has a living roof that makes it look like a hobbit's dwelling. We built the bathhouse on the site of an old sugarhouse, making its foundation out of the same rocks that a century ago held up the sweet steam of the maple arch. Around the bathhouse are huge craggy "wolf" maples that were once tapped for sugar and grown in open fields and are now sprouting fungi and starting to decay. Soon we will put up tents, assemble beds, and get the water flowing in the bathhouse, and by summer, this spot will house twenty-five people each week who are coming together around their work in land justice and social change. Everyone eats together in the upper story of the barn, enjoying food largely grown on the farm and prepared in the farmhouse. Just seeing the tent platforms makes me feel weak, thinking of everything that needs to be done, and hearing Peter's chainsaw just up the road, I know he's feeling the same urgency around preparing for the season. But there may be no more beautiful spot on the farm; this is the place where Peter and I put up a canvas tent and slept our first summer on the land, and where we were married. Today, on this most splendiferous spring day, this hill is the perfect place to be.

Wren runs ahead and claims a tent platform as her lookout kingdom. Within minutes she has shucked off her rubber boots and socks and is swinging on a wooden beam above the platform where in a couple of weeks we will stretch canvas. The wind frisks up the dry leaves of last autumn in tiny tornadoes like an invisible hand spinning tops, kicking up the pungent scent of dirt and decay. The two valleys drop away and rise up again in layers; the sky feels huge above and beneath us, as if we stood on the prow of a ship and could fall into the sky like water. As the sun comes over the forest edge, white light blots out the hills, splashes down through the new leaves, and blinds us.

Hot from the climb, the sheep stop and start to group together. Lobo, our guard llama, stretches his long elegant neck out to nibble tender new leaves from a cherry tree. The ewes stand for their lambs as they butt violently at their udders in thirst, one lamb from each side, then all the animals begin to bed down at the hill's rise, sides heaving and jaws moving in a circular rhythm. They have eating to do, but for now they rest, "placid and self-contained." Just one ewe, an older animal with a touch of mastitis, walks around picking out dry oak leaves near the woods. I wonder if there's a different, even stronger medicine in what has died, and I remember how farms in Scandinavia traditionally cut branches for leaf hay to feed the animals all winter. This ewe knows something I don't.

Wren finds a wide patch of wild strawberries in the sun. When the berries are ripe they are so fragrant that we follow our noses to them; they are "smelled before they are seen," wrote Thoreau. We flop down on our bellies amid the white and yellow flowers that are scattered brightly across the south-facing slope of bedrock, inviting the bees to visit them. Soon, we'll find the drooping red droplets of the first ripe berries, nestled in the grass and warm from the sun. These earliest of summer's gifts are about the size of a fingernail, speckled with miniature seeds, and tart on the tongue. There is nothing more delicious. Wren likes to look for "fatties," and saves up several to eat all at once by smushing them into her mouth—leaf caps and all—from the cup of her palm.

Nearby, I show her the diminutive plant called shepherd's purse, with its tiny white flowers that will become seeds the shape of a triangular leather bag shepherds carried under the arm. It's considered a women's herb—soothing to the menstrual cycle and said to help with bleeding after childbirth—but is also helpful for diarrhea. Another good plant for the sheep, I think. Next to it I spot one of those tiny thin-skinned snails, inside which is an invisible worm that can find its way into the brain of a sheep and drive it mad.

MARY COMES STRIDING UP THE HILL, CANVAS COAT OVER HER shoulder, sweat beading across her freckled nose under her glasses. She grins and throws her backpack down.

"Hey, no fair, I want those pajamas!" She crackles with energy. She admires Wren's crook and snaps a few photos of her posing with it on the platform, a huge badgerface ram, Merlin, resting in the shade behind.

"Okay, you can go now," Wren says after a while, much to my surprise. "I want to be a shepherd all on my lonesome." She has a collection of acorns on the platform and is sorting them into some kind of game with marks she has scratched on the wood.

"Just keep an eye on that guy," Mary says, pointing to Merlin. "The others like to follow him."

"We'll be just along this edge of the woods, so holler if you need us," I tell her.

I give Mary a pair of thick leather gloves and some wire cutters, and we head off. Mary, who is in her early twenties, is the first person I hired to help on the farm. (We hired others to help us with the retreats and had many people living with us each summer.) She called one day in the fall, saying she wanted to apprentice on a sheep farm. I didn't have apprentices, or hire labor in the fall, but she lived in town, so I invited her up. I was trying to get the ram lambs out of the field that day so they wouldn't breed their mothers and sisters, and it wasn't going well. I was pushing the last ram lamb through the gate when all the rams broke through a side fence and flowed right back into the ewe flock. I was swearing and yelling by the time Mary found me. I had forgotten she was coming and was in no mood to chat with someone about their curiosity about sheep. She didn't say a word, just started sheepdogging the field with me, bringing them all back into the chute like she knew what to do even though she was new to it all. She turned out to be fearless and strong but gentle with the animals.

I haven't met too many people who are more eager than I am to do the worst tasks on the farm. I've learned that Mary will do anything if she understands the reason behind it, or if it is an experience she's never had. Shortly after she started working we lost the most beautiful lamb of the season to barber pole worm. She was a large ewe lamb with the most stunning silky fleece, silver with black tips. Before I could bury her, Mary asked if she could save the hide. Something about carefully skinning the lamb that day led to her desire to become a butcher and to understand every part of our relationship to sheep. She is training to be an itinerant shearer and butcher and will go much deeper into this trade than I will, I already know that. But I'm honored that she started her sheep apprenticeship with me and even has the face of one of our rams tattooed on her arm.

We walk along a line of old maples and ash where barbed wire tacked into the trunks marks the old boundary of the field. The woodland air is cool and damp, and we kick up the scent of fungus, rot, and the metallic smell shared by earth and blood. The remains of a storm-felled tree covered with rich green moss crumbles into black humus laced with white fungal strands when my boot catches it. Behind it is a pocket of delicate wood hepatica, my mother's favorite spring wildflower.

The fence posts I'm looking for are long rotted, and most of the wire lies on the ground, buried in fallen logs and layers of moist soil and leaf duff, now and then passing through the heart of a tree that has sealed around it and enclosed the sharp wire in its flesh forever. The forest has meandered about thirty feet into the field from the place where farmers tacked this fence line sometime in the last century. As I show Mary where to find the barbed wire, I realize how much work it takes to maintain a field in this rainy, postglacial land with its deep seed bank in the soil; abandon an open space, and in under five years it will resemble a forest. All around us, the saplings and goldenrod and blackberry are evidence of this natural field succession.

It's probably safe to say that open fields are not a natural habitat in this part of New England at all. The meadows observed and recorded by early explorers such as Verrazzano in 1524 were maintained by Indigenous people to support game such as deer and turkey, to encourage the spread of grasses for basketry, clothing, and rugs, and to cultivate sun-loving crops like beans, corn, squash, and sunflowers. Our farm sits above the village of Waitsfield, chartered in 1782 and named after General Wait, who, with other militiamen quite certainly displaced a community of Abenaki in their managed fields and hunting grounds along the fertile Mad River. The Abenaki named the nearby Winooski River, "place of the wild onion," for a plant that grows in open woods or meadows along a floodplain.

THE FORTY-ODD ACRES OF CLEARED MEADOW OF OUR HILL farm were created by white settler Rufus Barrett in 1804. It's very possible that he did the laborious work of cutting down forests by hand with sheep—and profit—in mind. That year, as Lewis and Clark's voyage of discovery set west, an equally ambitious nation-building enterprise was getting underway in the east, and it was all about wool. The barely-viable hardscrabble nature of Vermont's homestead farms—and the entire landscape—changed drastically thanks to the arrival of one creature, the Merino, around 1810.

Merino sheep are a breed native to Spain, known for their exceptionally soft, dense wool. At the end of the eighteenth century some five million Merinos in Spain migrated from the villages to the Pyrenees each spring and back again in fall, herded by shepherds along public byways that were protected by the king (who taxed the wool) and required to be at least ninety meters wide. It was long believed that this transhumance—the journey itself—was responsible for the quality of their fleece. That and the salt put out on stones along the way.

Although the British Isles had been the epicenter of wool production since the Middle Ages, they did not raise Merino, and no other sheep had wool as fine. Spain guarded its breed closely for good reason, selling several million pounds of Merino fleece to England each year to make the best broadcloth to export around the world. Some of that cloth was sold to the colonies, accompanied by a stiff tax—a point not lost on Thomas Jefferson, George Washington, and Robert Livingston (one of the signers of the Declaration of Independence and a prominent New York judge), all of whom were preoccupied with how to raise sheep in America and produce their own wool.

Raising sheep in the colonies had not gone well, I learned. Not only did the colonists not have fine Merino wool, they barely had wool at all. The first sheep, brought mainly from England, quickly declined in health on the rough, just-cleared land; the wool was coarse and low in quantity. Wolves and mountain lions were constantly preying on the flocks, sheep were notoriously difficult to fence, and most villages had laws that animals had to be contained or the farmer would be fined. The first flocks of any size in the colonies were established on islands, where predators didn't live and fences weren't necessary.

Many of the founding fathers were sheep farmers themselves, working to improve the quality of wool. So important did they view the wool challenge that they included an appeal for farmers to grow their flocks for wool, not meat, in the Articles of Association at the First Continental Congress in 1774: "We will use our utmost endeavors to improve the breed of sheep, and increase their number to the greatest extent; and to that end, we will kill them as seldom as may be . . . and those of us, who . . . can conveniently spare any sheep, will dispose of them to our neighbors, especially to the poorer sort, on moderate terms."

Sheep in colonial America were both a source of ambition and embarrassment—poorly bred castoffs in ill health and yet essential to the cause of empire. The colonists' few sheep—as mangy and woebegone as they were—were so important that they even became a spoil of war. Knowing the revolutionaries' shortage of warm clothing and blankets for the winter, the British navy targeted sheep on Martha's Vineyard, stealing some nine thousand there and elsewhere along the coast in September 1778.

An opportunity came in 1801 when Thomas Jefferson appointed Robert Livingston minister of France. Both men saw in the beautiful French Merino flock at Rambouillet a breeding program they dreamed of for America, but they had to find a way to smuggle them out of Spain: "I hope to attain my object (more gradually indeed) by selecting two pair of the finest Merinos I could find, and sending them over under the care of one of my servants; believing that so small a shipment would not be noticed and intending to follow them by others," wrote Livingston. With this foursome, the first Merinos arrived in America in 1802, to be followed by 6,000 more in 1810, which Livingston smuggled through Lisbon as Napoleon invaded from France. Out of 6,000, only 610 of the stolen sheep survived the voyage to New York.

William Jarvis, American consul to Portugal at that time, also took advantage of the wartime chaos and acquired about 3,500 Merinos, with which he established a substantial sheep-raising operation in tiny Weathersfield, Vermont, soon to become the Merino capital of North America for a brief but feverish spell in the first half of the 1800s. By 1850, the state was producing over five million pounds of wool, with as many as 373 sheep per square mile. Merino rams were fetching as much as $10,000, an absolute fortune for a sheep even today. Sheep fever led to the growth of textile mills, a few Greek Revival mansions and local fortunes, and widespread devastation of the land.

MARY AND I MOVE ALONG, CUTTING THE BARBED WIRE FLUSH with the tree trunks with our fencing tools and then following all three spiky strands along to the next tree. Most of the time we are pulling the wire up from the ground, thick gloves protecting our hands. The oldest strands—probably put up for cows a century ago—are so brittle that they break as we pull, and we have to kick around in the leaf duff to find the lines again. We cut, pull the wires out into the blinding light of the field, then return to the dappled woods to find another section. It's satisfying to look back and know that we've cleared the remnants of something inherently violent—meant to look menacing and inflict pain—from this lovely fertile edge; the forest had been taking it back and burying it in its own time, but it feels better to remove the past's transgressions on this field.

Mary grew up in Vermont, and I just over the border in New Hampshire, and yet neither of us, we realize, has ever laid eyes on a Merino sheep. I've seen nineteenth-century engravings of animals with huge rolls of loose skin covered with fleece so dense that it resembles an out-of-control fungus, the sheep itself bigger than a barn. Mary asks me if this land was ever a Merino farm, and I don't know the answer. We have no record either way. Sheep were definitely plentiful in this valley though. Photographs of the mid-1800s show hillsides clear-cut right up to the ridgelines, zigzags of stone walls, sawmills—most of which are now gone—with piles of logs all along the Mad River. Anna Bixby Bragg—whose family farmed the hillside above our farm long enough that this part of town became known as Bragg Hill—wrote down a story for the centennial celebration of the town in 1898. The story had happened some years before, in 1826, in the heart of sheep fever. She told the story of a doe that was chased out of the woods near Shepard Brook straight into a clearing where a farmer named Carpenter was working. "Judge Carpenter caught it in his arms, and seven or eight hunters coming up just then, he told them that they could not have the doe, but that each one of them might go and select a sheep from his flock." This they refused, preferring to share a single deer. Sheep were so plentiful then and deer so few that no number of sheep could be traded for one sickly doe. Whether Judge Carpenter, out of a sense of justice, wanted to save the doe's life or have venison for dinner himself we will never know, but what is clear is that deer were precious, having been hunted almost to extinction in Vermont over a span of just a few years. If you saw a doe, you wrapped your arms around her and made bargains with armed men. Sheep, well, they were like locusts.

What happened to all those millions of sheep? How did they disappear so quickly and completely? My guess is that very few farmers ever solved the problem of poor land, poor fences, and predators successfully enough to keep sheep for more than a few years of quick cash. The sheer labor of the sheep enterprise in the Northeast must have seemed like madness once the endless western prairies opened. Vermont's sheep days were short-lived. By 1870, the price of wool had plummeted; Vermont farmers slaughtered their sheep by the thousands rather than feed them through another winter with no market for their wool. Many farmers headed west, with or without sheep, to find greener pastures for themselves, abandoning their houses to slump into the cellar holes and their devastated hillsides to grow over with trees.

One Vermont farmer, George Perkins Marsh, whose 1864 book *Man and Nature* is often called the founding document of the conservation movement, keenly noticed the changes in the land wrought by sheep fever. His account of its sudden collapse is one of the only primary sources I've ever found. "Within a few years, sheep have been killed in New England by whole flocks, for their pelts and suet alone, the flesh being thrown away," he wrote from his home in Woodstock, Vermont. And yet it was the land—which had been clear-cut for pasture more rapidly than at any other point in history before or since—that Perkins Marsh mourned most: "The felling of the woods has been attended with momentous consequences to the drainage of the soil, to the external configuration of its surface, and probably, also, to local climate." How prescient he was. When we did some work on our kitchen in the farmhouse, Peter found a water-warped copy of an 1895 *Yearbook of the United States Department of Agriculture* hidden inside the wall. One of the articles encouraged farmers to replant trees: "The forest waters the farm," it said, meaning that the forests act as a sink for rain that will nourish the crop fields. By the late nineteenth century, Vermont had seen the damage of its sheep greed and wanted its forests back.

In the last one hundred years, rural areas of Vermont and the rest of New England and New York have grown back to forest. Healing has happened. Populations of wild turkey, bobcat, lynx, bear, and white-tailed deer—all of which were locally extinct—have rebounded as their native habitat returned. But I believe there is also a place for the meadow. In Vermont's brief farming history, its open fields have become an iconic part of its identity, and they are important not just for aesthetic reasons, but also for ecological ones. History doubles back, but the cultural lens has changed. We have changed the world, not only in the micro—the plants and trees on this hillside—but in the macro: the climate and eco-balance of the globe. The drive to keep sheep during Merino fever was economic but also about nation-building; now, many sheep and grass farmers are equally motivated by eco-restoration in the face of climate change. If we graze sheep differently, in smaller numbers, they can be part of maintaining a diverse meadow ecosystem, restoring habitat for bobolink and Savannah sparrow and other field-nesting birds that are in decline, and for a whole universe of field-dwelling insects, reptiles, and other creatures I know little about. Grass farming can build soil health and deter erosion, and—by building extensive root mass below the soil—it may even play an important role in carbon sequestration.

MARY AND I HAVE FALLEN INTO A QUIET RHYTHM, JUST listening. As we stop to bundle wire, stepping out into the buzz and shimmer of the field, I realize there is another reason for keeping and considering the meadow, and it has to do with restoration of the spirit. To work outdoors in May is to feel like one small player in an immense symphony of sound and industry, and it is in meadows that I feel the most intense communion with the living, singing world. As my hands work, the rest of me is taking in, as if through my pores, all that surrounds me. Dozens of species of warblers and other songbirds are returning to the north country to nest and raise their young; thousands of bumble, honey, orchard, mason, and green bees hum among the early shadbush and hawthorn blossoms in the hedgerows; and spring peepers and wood frogs come out of the woods and mud to pulsate in the open water of the pond and field marshes.

Each May I also worry. I worry that the dawn chorus is quieter than in the past, the rites of spring less bright. I fret until I see the swallows and bluebirds return to their nest boxes near the blueberries, hear the red-winged blackbirds' wheeze and the bobolinks' joyful bubbling cadence in the grass. The spring peepers are loud and bawdy guests, singing and mating like mad right below our bedroom window. But out here in the field is a quieter courtship aria, an almost imperceptible chirrup of the earliest insects: "a minor Nation celebrates / Its unobtrusive mass," as Emily Dickinson wrote. It is hard to put my finger on diminishment, easy to convince myself I am imagining it, and yet easy to worry that I'm not. We have no trouble cataloging the loss of single species, but it is the diminishment of all species that will cause nature's intricately woven system to collapse, and diminishment is harder to perceive. As our experience of nature also dies, our senses no longer know how to measure the world around us. Can one sense the diminishment of all species if one has never been a quiet member inside the full symphony of spring?

It was in the spring grass that the theologian Thomas Berry had a boyhood epiphany of belonging, described in his essay "The Meadow Across the Creek": "The field was covered with white lilies rising above the thick grass. A magic moment, this experience gave to my life something that seems to explain my thinking at a more profound level than almost any other experience I can remember. It was not only the lilies. It was the singing of the crickets and the woodlands in the distance and the clouds in an otherwise clear sky. It was not something conscious that happened just then." Later in life, Berry realized that this moment in the meadow as a boy had defined his deepest beliefs: that everything in nature is an interconnected, complex community sharing its own story, that all things have "the right to exist and flourish in their ever-renewing seasonal expression." The universe is "a communion of subjects," not a collection of objects, he said. "We have silenced so many of those wonderful voices of the universe that once spoke to us of the grand mysteries of existence."

One extraordinary day, a year almost to the day before Wren was born, Sister Bernadette and Sister Gail from the Green Mountain Monastery brought Thomas Berry to our farm. We had met the sisters before, trespassing on this very field the same day that we were trespassing. They, like Peter and me, had put in a proposal for coming to this land, and they were sneaking a peek. They wanted to set up a sanctuary in Thomas Berry's tradition—an earth-based Catholicism he called Passionist, with the quest of being in sacred communion with the living world. (They later found land farther north.) Berry was eighty-nine years old when he visited us, tall and elegant and only slightly stooped, with kind, crinkled eyes. He was a person whose eloquence—but most of all, whose spiritual center of gravity—could realign how you move and talk, how you think. I wrote down some of what he said as closely as I could remember it in my journal that day. He talked about how from the last century until now we've made our home in the industrial age, rejecting nature as our place—ecology has become efficiency, nature has become natural resources. He told a story about talking to an old man reminiscing about his horses, how after running errands in town he would turn the buggy back home, finding the team had twice the energy. "We are the horses," Berry said that day in our farmhouse kitchen. "Someday we're going to run back to that place we belong and shake the harness loose and roll in the grass."

This land is teaching me about healing and transformation. It was damaged badly in previous assaults, by sheep, by clearcutting and erosion, by cattle. Those forces changed it, but with time it also became whole again. Even as I panic about the return of the birds from their wintering grounds and fear how other habitats have crashed around the world, I can also participate in healing an ecosystem here, most of all by looking around, noticing what nature already knows and is quietly teaching me if I learn how to listen.

MARY AND I HAVE MADE PROGRESS, ALL THE WAY DOWN THE hill to the southeast corner of the field where grows a massive apple tree in full flower, its ancient limbs curving down to the ground like the flying buttresses of a cathedral. This is a place Peter and I love to come with the girls. Inside the sanctuary of the tree it is cool, mossy, and dark, and you can climb and curl up in its strong horizontal arms. The south-facing cup of the hill, protected, and tilted just right to face the sun, must be the reason this tree has survived countless snowy winters and summer storms.

But for now, the bright morning has dimmed, and I can feel the wind grow from the north, treacherous and cold. Layers of dirty clouds like chalky eraser marks on a blackboard are blotting out the sun. We have pulled out a century's worth of barbed wire, unruly and sharp, snagging on everything and a devil to bundle up and carry. Our arms are covered in small cuts, and the side of my pants is torn away.

I have forgotten about lunch; it must be early afternoon. I wonder if Wren might have abandoned her post by now and gone to find Peter, but I can see the sheep still grazing above us and Wren's small form sitting on the crest of the hill.

"Let's call it," I yell to Mary, who is wrestling with a giant snarl of wire, trying to wind it around an old post to carry it up to the logging road.

"Want to go help Wren gather up, and I'll get the truck to haul this stuff down?" I ask Mary. She gives me a thumbs-up, and I head across the big meadow back to the farm. I break into a trot, which is what this hill always makes me want to do. I flush out a female bobolink—as brown and well camouflaged as her mate is flashy—from her nest hidden perfectly in the dead thatch of last summer. Once I found a wild turkey egg near here, all by itself and larger than my hand. I've tried to find bird nests in the field many times, but I've only ever done it by accident.

I return up the back way with the truck and see Mary and Wren herding the sheep out of the woods. "We just had a breakaway," Mary yells. "They decided to check out the woods all the way out to the stream, but they got super spooked in there. Maybe they were thirsty."

We agree that it's time to take them back to a fenced area, and Mary volunteers. It's the most fun job on the farm. I'll round up some lunch.

Wren comes trotting after me, shirtless and barefoot, chewing a piece of grass. Huge drops of rain begin to fall, opening like white poppies. She has always loved the rain and is drunk with spring. She runs over to a jutting piece of bedrock and jumps high, spinning her arms and legs out, falling, rolling downhill. Damp leaves of grass stick to her bare back.

"Mom!" she yells, bouncing up and racing ahead. "Come on! I'm starved!"

CHAPTER 5

STALKING COYOTES

We could hear ourselves breathe.

—JEAN GIONO, *THE SERPENT OF STARS*

After a long day, night falling at last on the summer solstice, we grab a blanket and walk out across the fields, east from the barn, past the garden and downhill toward the woods, toward the carcass.

The first fireflies flicker lazily ahead of us in the dusk-laden grass, and a barred owl calls from the forest's dark edge. A glimmer of daylight stolen by the river glints from the valley below. The night is humid, stars shimmering through a haze of heat shedding from the musky field, a half moon skidding above in a slick of oil.

In my thin tank top and running shorts, I am sticky with sweat. Hopefully the heat and mosquitoes will keep me awake.

My bare feet feel a hint of cool moisture in the sod.

"I think this is a good place," I say.

"Yeah, before it dips down and gets too wet. I put the body by the old well," Peter says.

He walks farther into the dark to check that the dead lamb is still where he left it, in a hollow by the cement cap someone built over a spring long ago. As the dark folds around me I see his headlamp bounce against the far trees. There, the remains of a stone wall covered with lichen and moss marks the boundary of the farm—a place I like to look for tracks and signs. I've seen foxes picking their way along this wall, a mother bear with her cubs snuffing blackberries, and once the flash of a fiercesome face. A fisher?

It's a fertile edge place, a rampart. A thicket of willows grows in the low ground, providing water and shelter for birds and wildlife. Predators intent on watching the spring for prey have the shelter of the trees dropping away behind them into a jumble of cliffs; there is a vague path over the barbed wire (yes, another place we haven't gotten to yet) and through a thick and silent cobwebby woods that ends in the village. We don't know that coyotes hunt here, but it seems a good guess.

Coyotes, like wolves, live in family groups of a dominant breeding pair, their pups, and several nonbreeding males and females. The family makes a den and claims a territory of up to ten miles around it, defending their hunting ground from other mating pairs who might try to settle. In an area like this valley, where domestic livestock provide a ready source of food, each coyote clan might defend a smaller territory. Though they are omnivores and survive primarily on small rodents, grubs, and fruit, coyotes will occasionally hunt larger animals such as deer, calves, and sheep, especially when they have a litter of pups to feed in spring and summer.

Coyotes are newcomers to the Northeast, newer than European colonists, newer even than domestic sheep. Given the dominance of their major predator, the wolf, in mountainous and forested regions of the north, the coyote found its niche in the grasslands and deserts of North America east of the Mississippi to the Pacific and extending south into Mexico. And there Coyote reigned. To dozens of Indigenous cultures across the Americas, the coyote—whose name comes from the Aztec *coyotl*—is a hero, a thunder bringer and dream maker as well as mad trickster, lusty buffoon, and moral guide.

By the late 1800s, in one of their many acts of unsettling America's natural balance, white men had hunted wolves and mountain lions to extinction in the East. They had never seen a coyote. Vermont's early farmers were able to trash their landscape with millions of sheep in part because they had wiped out most of the predators that might check their excess. Coyotes arrived east in the 1940s. There is genetic evidence that for several hundred years coyotes had interbred with wolves in the upper Midwest and Canada, and it is these larger migrant hybrids, with thicker fur and feet adapted to the snow, that we now call the eastern coyote.

Eastern coyotes live in our valley year-round. We hear them yipping and howling from the fields on bright moonlit nights, see their tracks across snow or mud, note their twisted rags of scat on warm rocks, and sometimes find a fragment of brindled fur on a wire. And yet they are so secretive that I've seen coyotes only three times: once when I tracked their singing at night and caught a glimpse of blurred shapes flowing across the road, once when I found a female dead with a bullet wound in the woods, and once while waiting to unload a ram for a neighbor—a magnificent gray-and-red coyote the size of a German shepherd watched me boldly from the field's edge. From watching coyotes on game cameras, I know they are incredibly playful and social—rolling over each other, wagging their tails, and licking in greeting—in nearly every way like the dogs we love.

I HAVE COME TO LIE IN THIS FIELD AT NIGHT NOT ONLY TO SEE a coyote but to kill one. The ribcage, spine, and head covered in a blood-soaked rag of white fleece—feet dangling like a marionette's—that now lies near the spring is what remains of a valuable yearling ram I had promised to a buyer. It is the latest sheep we have lost this summer, and now our bait.

My eyes accustomed to the dark, I stamp down a small area so our blanket will lie flat over the rough ground and tall weeds, space enough to lie and wait. I leave a fringe of goldenrod between us and the spring, parting the plants just enough to see through. I lick my finger and hold it up but detect no wind. We'll be lying about forty paces from the carcass, close enough for a coyote to smell us, wind or no wind. Their noses are as keen as a whale's sonar. But maybe the stink of rotting carcass will overpower our human stink. Or maybe the coyote will be hungry enough to risk it. This is the season when the pups are big and growing fast but still don't hunt for themselves, so the adults are driven by necessity.

More than half of the coyote pups in each birth litter typically don't survive past the first four months due to starvation and disease. The adults in the pack bring meat back to the den, either in pieces (a leg, a rib) or as offerings carried home in the belly that can be regurgitated. To take down a large animal, coyotes might work as a pack, but more often they travel and hunt singly, stalking small rodents and grasshoppers and raiding bird nests for eggs. When it comes to feeding on a carcass, the alpha pair will often police who in the pack gets to eat and how much.

I don't know why, exactly, but I anticipate the alpha female coming back to this carcass. She would still be lactating but free to leave her pups behind with yearling coyotes as caregivers at the den. Her mother's hunger would be insatiable.

Peter appears behind a pool of light, the Remington .22 that his father gave him under his arm, his hand fingering ammo in the pocket of his jeans. He peels off his cut-off rubber boots, unbuckles his belt and pants, and tosses them onto the blanket for a pillow, then stretches out his long, lean frame. Peter doesn't hunt now, but he grew up in the woods of Connecticut shooting squirrels and birds as a young feral child and was state pistol champion at the age of twelve. I've seen him take pigeons out of the air.

"It's still there. What's left of it. But they seem to have gotten a taste for lamb, so I bet they'll be back tonight," he says.

I take the flashlight and sweep its beam across the sheep enclosure near the barn below us. A hundred pairs of shining eyes turn toward me like tin lanterns flickering in the grass.

AFTER TEN YEARS OF TENDING A FLOCK, THIS SPRING I STARTED to feel like I knew what I was doing. I had started with eight sheep, and now I have close to ninety. The lambs are even-sized, growing fast, with luminous silky fleeces that crown their bright faces and ripple in the wind. There are lambs that stand out by their distinct beauty, glowing with health, their movements quick and confident. These are the animals I sell to people wanting to start a flock. A purebred Icelandic ram can fetch as much as $800, a ewe lamb $600, which seems like a whole lot of money for a sheep—close to prices Merino sheep farmers charged a hundred years ago, but not since—but even with these prices, buyers are contacting me from around the country.

Some of the fields clearly show the benefit of the sheep's rotational grazing and fertilizing. The white and red clover have come in without seeding, and the perennial grasses grow thicker and darker green. I look at the fields now and think, *There's a nice sward.* I relish the word but never felt privileged to use it to describe our thin mangy fields. *Sward* is from fourteenth-century English meaning skin or hide, the "greensward" of the earth being its sod, its sheepskin.

It's always a folly in farming to think you know what you're doing. Everything Peter and I were doing on the land took enormous energy and will. It was stressful and all-consuming. Each winter was a marathon of planning for the social justice projects and programs we would manifest through our nonprofit, which now had a board, a faculty of two dozen incredible people from around the country, and a sizable budget we needed to raise to continue to bring people to our programs tuition-free. Each year, we were also expanding the crops and animals on the farm. The summer, short and intense, was our reckoning—a lush, gorgeous, manic manifestation of all our plans, a season of celebration. But summer can also be a killer of dreams. Farming, I had learned, is a giant game of Jenga: everything that can happen in each move balances on everything we have built before and yet is made precarious by all the things we can't control, like the amount of rainfall or heat, new pests and diseases, parasite lifecycles in the soil, the strength and health of our own bodies, or the pack dynamics of coyotes.

Coyotes are so widely adapted that now one does not start a sheep flock anywhere in rural America without thinking about how to protect them from coyotes. Fences are not enough. And for most of us, bringing sheep into a barn each night is not practical. Also, in the hot summer months, sheep get most of their nutrition in the night, choosing to rest in the shade in the heat of the day, so it's important to keep them in the pasture. Bringing them back to a barn also exposes them to parasites and diseases that build up in concentrated areas. The beauty of having a grazing system is that the animals and the land are always working together, enhancing each other's health. But coyotes can disrupt all those plans.

To protect the flock from predators, most people choose a guard animal; large breeds of dogs like Maremmas and Great Pyrenees that were bred in Europe specifically to guard sheep are popular, but llamas, donkeys, yaks, and cattle can all work. What the flock needs is a guard with enough size, vigilance, and intimidation factor that the coyotes, who are by nature opportunists, will decide that it's too much work to go after a sheep and will continue hunting mice and eating carrion. It's essential the guard animal bonds with the flock and innately understands his job as protector of the smaller furry creatures who like to cluster around his legs. We have friends whose Maremma aggressively protected sheep from any intruder to the farm, including the mailman. With all the visitors we have, we decided not to risk getting an overzealous guard dog.

Until this summer, our guard llama, Lobo, had done a fine job of being that benevolent protector of lambs. But one day he got lame, then started staggering when he tried to run. A week or two later he lay down and, within days, wasted away to a heap of bones and velvet. Nerve damage and wasting disease are all symptoms of the meningeal worm carving its tortuous track up the spinal cord and into the brain, and it's heart-wrenching to witness. Llamas are particularly susceptible. They pick up the worm through those tiny, gleaming golden snails in the grass, deliverers of death.

On July 4 last year, Peter and I buried Lobo with the tractor, tipping his horse-sized body into a deep muddy trench that we scraped out with the bucket loader and two shovels in an afternoon of hot torrential rain. Lobo had guarded our flock from coyotes without fail for nine years, but in the winter, coyotes dug through the snow and frozen ground and dragged out his bones; it was as if they were warning us that they were now in charge. I noted the warning, but not enough, and by spring I still hadn't found another guard animal. I didn't know how bad it could get.

The first two lambs disappeared in the night from the birthing paddock. They were small enough for a coyote to snatch and carry in its mouth as it hopped the fence. A month later, the third lamb was still alive when I found it in the field. The mother, Europa, was standing near the fence line above a limp form: Her lamb was still small, born later than the others, and never strong. I walked over to the lamb, and it lifted its head weakly. She was mottled brown, white, black, and silver like her mother but specked with blood. I felt her body and parted her silky fleece with my fingers to see puncture wounds on either side of her ribs and neck: the mouth of a coyote. She would have been small enough to snatch but heavy enough that the coyote put her down at the fence edge, possibly scared off by Europa before it could drag the lamb underneath the wires.

I carried the lamb like a fragile scroll, hand cupped under its soft belly, arm extended so Europa could pick up the scent. She followed me in a nervous zigzag, strangely silent. At the barn I washed the lamb in a bucket with betadine soap and put her in a cool clean pen with her mother. A few days later the lamb was nursing well and seemed to have survived her brush with death.

I put the pair back with the flock, but it was too soon. I should have known. Her wounds begin to fester without my daily cleaning, and right away flies laid eggs on the scabs under the wool. The eggs hatched into tiny white maggots—thousands of maggots—and they started excavating and eating her flesh starting from a small wound. To see fly-strike is the worst thing you can imagine. Sometimes you can reverse the damage, and I tried. As I sprayed her with fly repellant and squeezed a liquid antiseptic over her body, the maggots rained down like grains of white rice. She had too many wounds, and she was being eaten alive.

I asked Peter to put the lamb out of her suffering and bury her in the woods, which he did very matter-of-factly. Peter helps me with birth, but more with death, and never complains. No matter how many deaths I see, I still take them hard and can easily blame myself. Peter, who isn't around the sheep as much, mostly feels sad for me.

The Trickster was getting the better of us. After Europa's lamb died, despite our putting up a double layer of electric fence at night, installing blinking red lights that were supposed to deter predators, and pitching a tent to sleep nearby, Coyote came in without a whisper and took a fourth lamb sleeping near the edge. Now I walk the fields with a constant feeling of dread; every rock and fallen branch in the distance looks like a carcass, and I find myself running to it to see which animal was taken.

Yesterday, after I found what was left of a yearling ram by following a raucous flock of crows, I felt such a murderous instinct against coyotes that I didn't hesitate when Peter suggested we hunt them ourselves. Coyote season is always open in Vermont. There is no limit to how many you can kill, or when you can do it, a law that seems innately wrong, though I hardly stopped to think more about it. I liked that it made things easy.

WE LIE ON OUR MOUND OF TUSSOCKED GRASS ABOVE THE hollow where water springs from the earth, our bait in front of us in the darkness, waiting.

We are on our bellies, arms crossed, cheeks resting on our hands so we can whisper, facing each other and pressed into the hot grass, dew drenching our bare legs and the seed heads of timothy and orchard grass dropping low to tickle our necks like the dance of a persistent and annoying fly. I feel something animate and stirring rise up from the hide of the earth to merge and mingle with my body, drawing me down and nestling me under the drenching silver of the sky.

We are like fishermen on the beach, baited lines running out into the waves to lure something wild and unseen, a hungry mouth from the deep. There is nothing to do but wait and watch, heart strained with hope for a sign, hands ready to act swiftly with any motion. We are watchers, we are hunters, we are snipers.

I notice that my excitement is tinged with something else, not fear or doubt . . . more like remorse. To hunt a predator feels like an act of malice, of dominance, of ego. We see ourselves in the predators of the wild; to eat a coyote would feel like an act of cannibalism. There may be no animal that has been more revered and reviled, romanticized and persecuted, adopted and murdered than the coyote.

I want so badly to put my head down, but I don't want to cover my ears, so I prop my chin on my hands, my nose pressed close to the bitter goldenrod crushed beneath the wool blanket, which has its own smell—like old books and sun, musty and comforting, drawing me into sleep.

The only part of me still alert is my ears, my listening muscles crouched and thrumming. I will my ears to stand watch even as I feel my brain's gears slip, thoughts mixing with the images of dreams. The electric throb of cicadas enters my veins like an ether. Beside me Peter is already gone, his body drained of all tension, his left hand on his rifle and his right touching my thigh, fingers twitching as he drifts away.

Time passes, an hour or four, I have no idea. Then, my ears pick up the faintest rustle in the grass from the hollow. I'm instantly alert. A pair of emerald eyes stare at me from what seems like a few yards away.

Later, I will remember the look in those eyes as sad, recusing: "You tricked me," they say. I will think of Aldo Leopold, who wrote of killing a mother wolf, and how watching "a fierce green fire" die in her eyes was the awakening of his ecological consciousness: "I was young then, and full of trigger-itch; I thought that because fewer wolves meant more deer, that no wolves would mean hunters' paradise. But after seeing the green fire die, I sensed that neither the wolf nor the mountain agreed with such a view."

I touch Peter's arm, and in less than a second, he has his rifle to his shoulder, aims at those steady green points of light, and fires. Between the pull of the trigger and the sharp crack of the gun, she has vanished.

BEYOND WHERE WE LAY THAT NIGHT IN THE FIELD IS AN OLD apple tree, nearly as old as the field itself: from 1804. It stands alone, perhaps sprouted from a seed dropped from a long-gone fencepost by a long-gone blue jay, from a seed with colonial parentage, from an apple that ripened when Thomas Jefferson was president and Robert Livingston was negotiating the Louisiana Purchase and plotting to smuggle the first sheep out of Spain.

This tree's apples are small and golden, their skin with the roughness and earthy patina of unglazed clay. The tree's original trunk split open years ago and lies splayed on the ground, massive branching arms hugging the earth and simultaneously reaching for the sky with a thousand water sprouts—tree as fountain, tree as mother.

One can duck inside this cool, mossy green fountain to sit on snaking limbs and breathe in the humid air of rainforests. This tree has such undeniable presence and resilience that Peter and I, as well as others who have lived and worked at the farm, have been drawn to its sanctuary to talk about difficult things, to heal wounds, to ask for forgiveness. On the half-rotten bark of one of the open arms of the tree, someone has placed a small metal statue of the Hindu goddess Tara. Born of the suffering world, Tara is often evoked for guidance and protection, for guardianship.

Tara wasn't protecting us that summer, but she stood as a reminder of the ancient tradition of humankind turning difficulties into deities or, to put it another way, of worshipping deities that have within them both creative and destructive forces, anger and violence as well as love and protection. "In man, so in the god," my father once said, and it stuck with me. Spirituality is born of the need to understand suffering, not remove it.

Thinking about the importance of Coyote as mythic figure, as a deity for many Indigenous peoples, helped me accept that something larger than I understood was at play. It makes sense that our deities are manifested in times of trouble, not so much to protect us as to uplift and hold at a distance something out of our control. In the stories of Indigenous peoples all through the Americas, Coyote teaches about humility, about the impulsiveness of our desires, about knowing our place in the order of things; and his naughtiness is told in a way to make people laugh. This all gives relief to the human suffering.

I thought about Lewis Hyde's premise in his book about the coyote, *Trickster Makes This World*: "The origins, liveliness, and durability of cultures require that there be space for figures whose function is to uncover and disrupt the very things that cultures are based on." Native people wisely incorporate into their cultures this figure of the disrupter from the natural world, where they see the same things at work in their own kind; to stay vibrant, they understand, nature and human culture both require forces that constantly interrupt and rebalance the order of things.

After the hunt, with a big problem still on our hands, I decided to learn everything I could about coyotes. Our neighbors Ky and Lisa are wildlife biologists who, for the past fifteen years, have spent every winter studying wolves in Yellowstone or Isle Royale National Park. What I learned from them, and from reading papers they gave me, is that we, with the other farmers in the valley, had probably created the predation problem ourselves. Only by understanding more about them might I be able to shift the dynamic so that the coyotes changed their hunting habits and we could peacefully coexist.

Coyotes, because they have been colonized so long by the more dominant wolf, are incredibly adaptable in the face of pressure. Hunting coyotes can simply disrupt the dominant hierarchy of the pack, fracturing it so that more adults begin hunting to support themselves. Killing of alpha breeding pairs also prompts younger or less dominant animals in the pack to start breeding, leading to more packs and a greater population. With more pups, the coyotes are under more pressure to find larger sources of food. And by leaving young sheep and cattle that had died of other causes in the woods and on compost piles around the valley, we farmers were supporting these larger coyote litters all on our own.

In the West, almost as soon as they encountered the "prairie wolf"—as Meriwether Lewis called his first sighting in Nebraska in 1804—white settlers were intent on hunting and poisoning every last one. This zeal was much more about a European ideal of a predator-free pastoral landscape for livestock and a game park for the uncontested human predator than it was about necessity.

In the killing fields of the Great Plains in the 1840s, white settlers, intent on wiping out Native Americans and their sacred buffalo, were lacing dead buffalo with strychnine—a cheap unregulated poison made from the seeds of a tropical nut tree—to kill scavengers by paralysis within minutes of touching the meat. States started offering bounties of "$1 per scalp" (in Kansas in 1877; word choice very intentional) on wolves and coyotes. In the opening of the twentieth century, the new Biological Survey, with an original stated purpose to catalog species, soon became the first government agency with a mission to clear predators for ranching interests, appropriating funds and accepting private donations for this purpose. Biological warfare—poisoned bait stations—was their main tactic. When the smart coyotes learned to be "bait-shy," the rangers changed their lethal cocktail and laced carcasses with thallium sulfate and sodium fluoroacetate (also called 1080), poisons that act more slowly, making coyotes blind and naked, in the case of thallium, or making them convulse and eventually stop breathing hours later, in the case of 1080.

Though poisons were banned in 1972, 1080 is still used today by Wildlife Services on large Western ranges in the form of sheep collars that eject poison when bitten by a predator—essentially protecting one's meat by locking a canister of poison around it. Shepherds and their dogs guard sheep flocks on the vast ranges of the West, and predation has been, and still is, a problem. But over the decades, it seems very likely that millions more dollars have been spent on bounty hunting, poisoning, intentionally introducing diseases like sarcoptic mange, aerial hunting, and even sterilization of female coyotes than has been lost in sheep and cattle revenue.

Wolves were hunted and poisoned out of every state in America by the 1920s, but the coyote hung on. Montana paid bounty on 111,545 wolves and 886,367 coyotes between 1883 and 1928. By 1920 the wolf population in Montana and elsewhere had dropped almost to zero, but coyote bounties remained consistent, about 30,000 per year. How did they survive?

With the exception of Adolph and Olaus Murie, who studied coyotes in Yellowstone National Park in the first half of the twentieth century, comparatively little effort has been made to understand coyote behavior and their role in the ecosystem, or to try nonlethal methods of protection for livestock and test coexistence. One contemporary biologist who has dedicated himself to understanding coyotes is Robert Crabtree. In a scientific opinion letter posted from the Yellowstone Ecological Research Center where he works, he wrote, "This letter also addresses a long-held belief that human control of coyote populations are 'necessary,' similar to 'mowing a lawn' to keep it from growing out of control. This belief has no scientific basis whatsoever." He goes on to explain why hunting coyotes does not lessen pressure on livestock, and how coyote populations stabilize in balance with their environment. I learned that coyotes adapt to stressful conditions, as do humans, with "fission-fusion" behavior, meaning that they bond and cooperate with others when it benefits them but also resort to competitive, individualistic behavior when that is more advantageous to their survival. Killing coyotes actually increases overall predation, because in response to their social order being disrupted and their survival threatened, the remaining coyotes in the pack will increase births and hunt more opportunistically.

The more I have learned, the more respect I have for these resilient, intelligent members of our ecosystem who are still here despite extreme efforts in the last hundred years to entirely wipe them out.

Likewise, in the oral and mythological traditions, the figure of Coyote has escaped capture, assimilation, and scholarship. He is alive and well, ever shape-shifting and evolving, in the stories told by Indigenous elders across dozens of Native American tribes. This resilience gives me hope, and I no longer want to be part of an effort of domination and diminishment. I now see hunting coyotes as the act of colonization it is.

On a farm, nature gives us everything we have. It's the genesis and genius of every seed, every particle of humus, every mycorrhizal strand, every fruit-bearing tree. It seems only right that we harvest some and give some back. The lambs that go back to the earth through the belly of a coyote, or the countless blueberries eaten by a cedar waxwing, are seen as "crop loss," "damage," or "failure." I am learning to appreciate them as return, as gift, as exchange. The taker of the lamb can be the deity, the trickster who is also the thunder maker and brings the rain, the balancer of life and death that helps us see our humility and purpose as one cog in a greater, infinitely wiser wheel. It's all a matter of perspective. And perspective may be everything if we are going to turn around the path of destruction our belief of dominance has set us on.

We never shot a coyote. Instead, we adapted, and things improved. We stopped burying sheep on the farm.

We installed a dog-proof fence around the lambing paddock. We got another guard llama, Habibi—a magnificent, nervous animal with huge eyes and long ears who has the instinct to herd the sheep to the highest point in every pasture at dusk—and an English shepherd pup, Rue. Now Rue is a year old and sees herself as ruler of our hillside. The coyotes seem to respect that it is her territory now. She patrols a circle around the fields each dusk and dawn, her nose to the ground on the traces of nocturnal footprints, hackles raised. If she hears coyotes in the distance she sits down, throws her head back, and raises her voice with them—long, low, moaning howls and high yips, her wild relatives' voices infinitely more complex and layered than hers but in total kinship. It is the wild and domestic sharing a song, and it is a beautiful sound.

CHAPTER 6

LIGHTNING

> The herds climb into the thorns and the furnaces of dust.
>
> —JEAN GIONO, *THE SERPENT OF STARS*

THE WEATHER REPORT IS DIRE: "Strong winds, lightning, and damaging hail, and up to an inch of rain an hour with flash flooding likely." These are the kind of summer storms we often get now—no rain for weeks, and then a deluge that carves deep gullies down our roads and overflows the rivers. This year farmers have lost crops and whole fields to flooding in every season of the year, the bloated rivers belching up contaminated silt and trash and trunks of trees. Changes in temperature as much as forty degrees in a day were not uncommon this spring, with one storm front crashing into another and reversing the season from winter to summer, then back to winter, within a few hours. My apple trees that started budding in February and flowered in April were frozen in May. I know it's crazy to make blanket statements about climate, and weather in New England has always been unpredictable, but when you farm, you count on certain things happening the same way year after year, at more or less the same time. If it warms too soon and the fruit trees blossom and then freeze, there's no fruit. If the river overflows its banks on a full crop in July, that's a disaster. The seasonal patterns are the floor beneath the game of Jenga. Even when other parts of farming were precarious, these patterns were the solid ground we counted on. But not anymore.

When I was a child, summer weather still followed a predictable pattern. The clear sunny days would get progressively hotter, the air getting heavier each day, and then thunder would start rumbling over the forest. Lightning would flicker on the ridge, and the thunder would come closer, louder, until the storm broke in heavy silver sheets over the green fields. Very suddenly, it was cool again. The cool air would slipstream under the storm like light under a door. For a few days the sky was an infinite crisp blue. Those were haying days, when we raced to get the grass cut and dried and baled before the next storm cycle came. We could count on a rhythm. Now, not so much. Last summer was all drought, and this summer is a jumble of extremes—scorching hot this May and June, and now day after day of monsoon-like rain.

I NEED TO WALK UP TO THE DAY PASTURE—THE PASTURE WE named after Ann Day—to fetch the sheep off the hill and into a more sheltered place. The portable fences can blow over in such storms, and even though the sheep will probably be fine, I am worried. If they get spooked and run out of their fence, they could be a long way from the farm before I know they are gone, and even if they don't, I'll be lying in bed tonight imagining it. Last night I had a dream that I was taking care of someone else's flock in a huge maze of corrals and pens, where rams and ewes of all ages were mixed and fighting, with bloody, broken faces and bare patches of skin. There was a horse in there too, dragging a saddle, and crows fighting over something dead. My dreams don't leave me alone.

I recruit Wren to go with me; she likes the role of sheepdog. It's a long walk back through the open fields with plenty of places for the sheep to run off, and I'm tired from crisscrossing the farm all day. Wren, who went back to public school after two years of homeschooling, has become a strong cross-country runner in middle school. I need her quick feet.

As we start up the hill the air is murderous, the sun like hot breath scorching our arms and necks. Heat clings to the hills like a silk nightgown to the body after a shower—shimmering, uncomfortable, expectant. My throat is parched, my temples pounding. I know I'm dehydrated from the day. The marsh grasses and field brambles cut our bare shins with their fine teeth, and the wounds sting with our sweat as deerflies buzz around our heads, leaving welts on our shoulders. Stuck to our thighs are bright-yellow buttercup petals shaped like teardrops, grass seeds like itchy grains of sand, and the little land snails that I pop with my fingers like roasted pumpkin seeds. We never saw land snails as kids and had never heard of a tick, or Lyme disease, or an Asian spotted-winged fruit fly that is now like a locust plague in our blueberries, puncturing the berries so that they fall to the ground. With the warming summers, new pests that have no native predators are surging north.

"It feels apocalyptic out here," I say.

Wren takes a swing at a huge horsefly on my back.

"Why does the air look orange like that?"

"Wild fires in Canada I've heard."

Fires in Quebec this summer are so hot and widespread that they are essentially unfightable. Our skies have had a brown cast for weeks now, and the wind smells singed.

After a pause, Wren says, "I basically feel like I've been preparing for an apocalypse my whole life."

"Really? Say more."

"I mean, most kids my age have been hearing about climate change since we were born and walk around with a sense of doom, knowing this world will be totally different before we are old. The crazy thing is that it sounds slightly thrilling to me. No rules! I like the idea of having to use my skills to survive."

This is not something we have talked about before, and I'm not sure what to think. Part of me wants to laugh. I think she's been reading too much cli-fi or imagining *Parable of the Sower*, which is one of her favorite novels, but we all know Octavia Butler is a prophet. I already see the changes and worry about what's coming. More violent storms, floods and pests will make it harder and harder to grow food even here, and this is one of the places people will want to live given that most of the South will be too hot and the West on fire. We will need to grow more food but may also lose a lot more farmland to housing, which we will also need. Wren may be right that having the skills to grow your own food in small plots will be essential.

As a family we have talked about how, for Indigenous people here in the Americas and elsewhere, the apocalypse already happened. For enslaved Africans it already happened. In Gaza and Sudan, as I write this, it is happening now. For disempowered populations on marginal lands and in toxic waste sites all over the world it is already happening. For us, for now, we have the privilege of imagining it. I don't remind Wren of her privilege in this moment; she has grown up steeped in conversations about this and the responsibility that comes of it. I am thankful for the moment that resilience and energy come to her instead of the fear and anxiety that come to most of her peers, and I want her to hold on to this. This will count for so much. Resilience is something she has learned from being out here, as participant in and witness to nature's constant cycle of life into death into life.

WE CAN HEAR ONE LAMB CALLING OVER AND OVER AS WE walk up the hill—a lamb because the voice is higher, thinner; like a child, one can tell. I'm happy it's a lamb calling. A mother calling urgently is more serious. At this stage in the summer the lambs are still small enough to get worried if they don't see mom nearby, but mothers often won't answer their call, letting them find their way back on their own, content to eat and have some peace from nursing. But every few hours the mothers do a check around for their babes, and if they can't find them they will begin to call loudly, panic in their voice. I recognize that panic because I've felt it myself.

We turn up the logging road and through a break in a stone wall to the Day pasture—sixteen acres of marshy grassland with a drainage running through the center from the steep craggy hillside above. After being our neighbor up here for ten years, Ann moved to a retirement home in her late eighties.

The sheep are being tormented by deerflies too, and they stand at the edge of the trees, heads down, ears flicking and feet stamping irritably. The lambs pant in their huge summer sweaters, small tongues out, sides heaving rapidly. Habibi, the llama, pins back his long ears and crowds under a honeysuckle, cracking branches, a look of intense foul humor on his face. Three goldfinches cling sideways to the very tops of the slender grasses, eating the seed heads, swaying slightly and then darting away, as if they, too, are in a hurry before the storm.

The animals begin to walk our way, as if expecting us. One lamb, looking ridiculous, trails an enormous blackberry cane that's tangled in her fleece. She doesn't understand why one of her sides is getting snagged by the grass and keeps hopping sideways to look behind in alarm. It makes us laugh. A worried brown lamb is calling for his mother from the trees, and I spot her grazing about fifty yards away, but for him she is hidden by tall grass, and he's too afraid to wander. As the sheep converge toward us, his mother finally responds to him, and he runs to her, butting his head under her for milk with such vigor and injured pride that he nearly lifts her back legs off the ground.

With the flock comes the smell of dung and dust, of sweat and sun, of crushed bracken fern like bitter tea on the tongue, a smell as complex and storied as the briny sea. An assertive black mouflon ewe with her two white ewe lambs leads at the front. She has been feeding them all spring with one good teat, her other one blocked with scar tissue from an injury. Her lambs look good, though one twin is much larger than the other by virtue of their nature. The larger one is more aggressive, better at beating her sister to the one good side of the udder. Because nature is raising these animals without much intervention from us, the lambs that are clever and full of gusto soon outpace the rest. Over the past fifteen years, I've seen how the flock has gotten stronger by virtue of who adapts the best to all the challenges of changing weather, diet, parasites, and disease. Those who adapt the best pass on that knowledge, or DNA, or both, to their young. Yet how can any species adapt fast enough now when the forces pushing that adaptation are constantly changing too? Adaptation takes time. Make change too rapidly, and it's the definition of catastrophe.

Wren knows to wait until every sheep has gathered, then she opens two sections of electric netting, and the sheep stream out like water from a lock. Their bodies crowd and jostle along the woods path and then form a single line, nose to tail, up the logging road. They cross a small meadow, then flow through a break in another stone wall in tall trees. The air is beginning to whirl and prickle with electricity, the leaves above us blurry and shattered.

Four ewes and their lambs veer left to a wild apple tree to look for early drops in the grass. They know this tree and make a beeline. Wren turns to yell something back to me, but her words are taken by the wind, which has risen to a shriek in the branches overhead. We don't have far to go, but our sheep are all splintering out now, and we are on the chase, as instinctive as well-trained dogs. I run uphill to make a wide circle around the trees to head them down, and Wren sprints out to the open field to keep them from breaking too far in front of me. I have to run hard to get ahead of the black mouflon who is booking toward the woods. If I can get close enough that she sees me on her flank and turns her head right, then she'll go toward Wren, and the rest of the flock will follow. Then I'll run behind, and Wren will keep them from going straight down the steep slope that leads all the way to town.

A black sky cloaks the shoulders of Mount Abe, making the mountain huge and ominous above us. Storm clouds build there, gather force, then fall along the ridge before coming across Mill Brook and up our high fields. Lightning flashes in a deep pocket of the sky, and with the next crack of thunder I know the storm will break over us before we make it back to the farm, not in drops but in one blinding rush carried by the wind. It feels good to run hard in the electric air, bracing for the rain.

We get the sheep through the gate by the barn and walk onto the house porch utterly drenched, as if we had jumped into the pond fully clothed. Peter comes out of the house, and the screen door tears out of his hand to crash against the wall. He is saying something, but we can't hear. He looks distressed; I know he had a grant deadline, and probably we have lost power. Our hair is plastered to our faces, and our rubber boots are full of muddy water and make loud sucking noises as we sit on the porch and pull out our bare grimy feet.

WREN HAS GROWN UP LEARNING TO READ THE LAND AS WE move about, thinking with sheep. The shade of trees or the aspect of the slope changes the moisture of the ground as well as the temperature of the air, the heat sink of rocks, and the grooves of old streams. The kinds of plants that grow in each place tell the story of that soil, its underlying bedrock, its aspect of sun and shade. All of these details tell where the grazing will be good or poor at any given time of the year. If Wren were to grow up and work as a shepherd in the Sierras or the Pyrenees, I believe she would quickly learn to see the land there, too, much as a trained musician learns a new instrument.

As shepherds, *we walk across.*

Transhumance is such an awkward word for something as evocative as "to walk across the earth," a phrase that conjures the long journey, the serac on the glacier, the saddle between mountains. But I like its highly adaptive meaning: Transhumance is the seasonal movement of domesticated herds based on the birthing time of animals and the life cycle of plants. In the Northern Hemisphere, this meant moving your animals from the low, sheltered valleys in winter to the high alpine meadows in summer. Spring comes first to the valleys, and the green shoots are nourished by the river and the silt of floods. This is the realm of gardens and orchards and cultivated fields. The low country is also the winter fold for sheep to eat dried hay and gestate their lambs when the hills are dormant and cloaked in snow. But as the earth tilts into summer and the sun bakes the valleys, it is the cool north slopes of the mountains that hold the moisture from the slow melt, shady alpine meadows and open groves of wind-hardened trees that have the best grass for the growing lambs.

Each land has its pattern language, written in scales large and very small, mapping the movement of weather and the microclimates created within its folds. I like to think of transhumance as the reading of a particular land's pattern language by the animals and tenders of animals who walk across it, practiced on every continent by pastoralists for thousands of years. It feels exciting to me to be part of that lineage and to be learning that kind of fluency in the language of my own home ground. At least for as long as that is still possible.

The phrase "pattern language" comes from Christopher Alexander and colleagues, from their book of the same name. They use the term to describe the patterns inherent in building and designing structures, but I think it applies just as well to pastoralism. When I read Alexander's definition of a pattern language, I thought of my study of pastures: "It says that when you build a thing you cannot merely build that thing in isolation, but must also repair the world around it, and within it, so that the larger world at that one place becomes more coherent, and more whole; and the thing which you make takes its place in the web of nature, as you make it."

The thing which you make takes its place in the web of nature, as you make it. This reminds me of something Peter always says about the land ethic of "leave no trace": While acknowledging the great damage humans have done to land, this ethic also ignores and displaces many peoples and cultures who have coexisted with and nurtured the land, who themselves have known how to make that place *more coherent, and more whole.* "The goal should be 'leave a beautiful trace,'" he says. Conservation is one of the dominant forces, along with development and drought and war, that has killed pastoralism as a way of life in many parts of the world. As we lose constancy in our earth's systems, we are simultaneously losing those peoples who have stored the most ecological knowledge and who have survived by using that knowledge to help the land adapt to change. Each pastoralist culture's map of the land, a complex and layered understanding that has evolved over generations unique to that place, is being lost. With it, the place itself will also be lost.

THE TWO PARTS OF THE WORD *TRANS-HUMANCE* TRANSLATE AS to "cross" the "ground," to make passage on the *humus*, the sacred earth mother of *human*-kind, words connected by the same root. In the Bible, *Adam*, meaning red, shares etymology with *adamah*, meaning earth, or red earth. We are made of earth. *Abel*, the name of the first shepherd, has the same root word as breath. To discover such entwined etymology in the English language feels like unearthing a long-buried cosmology, a lost treasure so encrusted and corrupted that we no longer recognize it as relevant. But pastoral and indigenous cultures, to which we all belonged at one point in history, know otherwise. As these traditional pastoral ways and the ecological knowledge they hold are tragically disappearing, teaching ourselves—however we can, wherever we live—to hold on to a pattern language of how to breathe and walk on this earth seems to have more relevance than ever.

I woke from a dream once with a phrase ringing loudly in my head: "The land is losing its memory." Losing its way of knowing how to do things. Of knowing how to adapt. This has stayed with me. The dream reminded me to start noticing more, observing more, and doing less. I want to speak the language from which these memories of resilience were made.

THE MORNING AFTER THE STORM, I REALIZE I AM MISSING a ewe.

Lately I'd noticed that this ewe, who was only a year old, was getting a small udder, and her sides were extra broad. She must have gotten pregnant way out of season, months after all the others, and later than I thought was even possible, given that Icelandic ewes come into estrus only in the fall. Another cycle out of whack. Ewes will often wander away from the rest of the flock to give birth, so I look for her everywhere around the paddock where we penned them up in the storm—behind the manure spreader and the old truck, by the wood pile and the chicken coop. Nothing. We must have left her behind.

Only two other times that I can remember have I lost a single sheep. They hate to be away from the flock alone, so if you can't hear them calling and making their way back, it is a bad sign. They could be well and truly lost, like the ewe who found our flock one day after having gone missing from her home farm for three weeks. She appeared out of the forest, disheveled and emaciated. She walked the boundary of the field I had just fenced, wanting to be with our flock but too nervous to let me herd her to the gate, skittish as a wild thing but determined to stay with us. I finally got her in with trickery, making a kind of weir with net fencing that funneled her in toward my sheep before they could run out. After asking around, I found out she had escaped from new owners, run many miles through the woods, crossed the Mad River, and found us, somehow surviving many nights alone. She'd been trying to find her old flock.

Early the next morning after the storm, I head out. The air is still thick and hot. The grass is smashed down and strewn with leaves and branches from the storm. My boots squelch through the standing water in the low fields as I make my way up to higher ground, the strong smell of field mint rising as I crush it into the mud beneath my feet.

I have other things to do, and I'm still looking for this animal. In my irritable mood, I think of something James Rebanks said in *The Shepherd's Life* about how a sheep could die out there just because we decide not to be bothered with it. I think of his three rules of shepherding: 1) It's not about you, it's about the sheep and the land. 2) You can't win sometimes. 3) Shut up and go and do the work.

I step through the stone wall into the Day pasture. I walk along the stone wall, looking for my ewe who might right now be in labor or too preoccupied with newborn lambs to make it back to the farm. A ewe wants to stay in the spot where her water has broken to give birth; she is now bonded to that piece of ground. If I want to move her to a safer or cleaner place, she will run back to that place in the pasture if she can. Once she has given birth, she is also reluctant to move until she is sure the lamb is sturdy enough to follow. The ewe will go without food and water for a day if she has to.

I see her from a long way away. She looks huge, lying on her side. I can tell—even from a distance—from the way she's lying that she is not in labor. She is dead. She might have "cast," meaning that she got stuck on her back and couldn't get up again and literally pressed the air out of her lungs. This is more likely to happen with a very pregnant ewe, and her unborn lambs will die inside her.

This ewe is bloated, like a balloon animal blown up at a fair, and there's bloody foam at her nose and mouth. The stretched skin on her belly is bruised. She looks like a cow I once saw that had been struck by lightning. My sister, who is a large-animal vet in the town where we grew up, says this happens more than you might think.

I walk a few steps closer. A ragged brown form moves out of the shadow of the dead ewe's belly. At first I think it's a weasel or rat, feeding there. I recoil, but it makes a small plaintive cry. I see that it's a lamb, drenched with sadness. It must have been born last night and stayed next to its mother all this time, taking shelter from the storm in the crypt of her body.

The lamb is utterly trusting and lets me pick it up. It's wet and tiny, honey-colored on the top and cream on its bib and belly, legs dark brown and hooves still soft like shiny patent leather. It curls its lips back and trembles, shaking its head, trying to suck on my sleeve. I carry the tiny thing back to the barn and hold it at the water trough. She drinks intensely, on and on, like nothing I've ever seen before. She is dehydrated and yet not at all weak. Usually lambs that have been orphaned or rejected are so weak they cannot suck at all.

"I have good news and bad news," I say to Kyla, who is working at the farm that day and comes around the barn, looking for me. "Here's the good news!"

Kyla works at the local high school during the school year and, for the past three years, has helped me with the farm stand, blueberries, and gardens in the summer. She's petite, with fine brown hair always pulled back, and dresses in layers of soft cotton and shawls in lacy patterns she knits herself. She's possibly the most sensitive, nurturing person I've ever worked with, and also the most devoted.

She's the kind of person who will love having an orphan lamb to spoil.

"No way! That is the most beautiful creature I've ever seen," she says as I tell her the story and hand the baby over to her. "It looks just like a deer."

She cradles the lamb, who is suddenly limp with fatigue, in both arms and heads inside the house to mix it a bottle.

I'm not sure what to feel—distressed or lucky. I have gained a surprise birth, but I didn't expect to lose this beautiful new mother, who would have had ten or more years of birthing ahead of her. I lost a valuable animal and gained a whole lot of work; having to keep this lamb alive by feeding it milk every few hours was not on the list. I know many farms take offspring away from the mothers at a very young age and feed them milk replacer and fortified grain in order to keep the milk and control the animals' nutrition, but I am not a fan of this system. I don't like bottle babies. Intuitively, it feels all wrong. Ruminants are not designed to eat a lot of grain; they are meant to eat grass. Nor are humans meant to eat a lot of grain-fed meat. Eating a small amount of meat produced by animals who are grazing less arable land might be the most sustainable food decision we could make. Most pastoralists around the world are producing meat and milk on land too poor to sustain cultivated crops.

I take the tractor to load the mother's body into the bucket loader so I can take her into the woods to bury her. The storm has washed out vast parts of the road, exposing large loose rocks that make the tractor lurch from side to side. All the gravel and sand has been washed into the ditches and clogs the culvert. Other things not on the to-do list: regrading the road and re-cutting water bars before the next storm.

I turn into the field, lurch across the foundation of the stone wall and into the long grass. As I drive the tractor down toward the ewe's bloated body, a brown lamb comes running full speed from the other side of the field, crying in a desperate voice much larger than its tiny form seems capable of making.

What the hell? I think as I stop the tractor, jam on the parking brake, and step down to pick up the lamb. Has it escaped out of Kyla's care and followed me all the way up the field behind the tractor? For several minutes I'm confused, but this one is a ram lamb, and I swear the other one was a ewe. This brother must have decided to run out to call 911, leaving his sister by the mother's side. What's incredible is how this small creature, who looks nearly identical to a whitetail fawn, is acting domesticated from birth. How does he know that a human is safety, is food, is the next best thing to his mother, who he must have figured out is never going to respond?

I put the lamb down, and he paws at his mother's stretched and blueish belly, tries to tug at her swollen upended teat. I wonder how long it took the milk to congeal after she died, or how long after they were born she was struck down. Somehow these lambs seem to have gotten some nutrition and are fighting strong.

I heave the ewe into the tractor bucket, lifting her two legs at a time and shuffling her in sideways. She's heavy and as stiff as lumber except for her head, which lolls and leaks a foul-smelling bloody serum. I scoop the little lamb up and put him on my lap, driving with one hand along the bumpy road back to the barn. He doesn't struggle. He seems to find the tractor seat and the noise and heat of the engine perfectly ordinary. In front of us, I can see his mother's stiff feet jouncing around with each bump in the road. I'm going to have to drive through the farm like this in order to drop the lamb off before I head to the forest to bury her. I hope no one visiting gets a shock.

Born domesticated, these lambs will need me to survive, and yet I have no desire for a pet. I will have to get up every other hour all night long and give them milk. I will have to sleep downstairs so as not to wake everyone else, and the dogs will want to be with the lambs, and the lambs will cry all night if they are in the house, but leaving them in the barn means making a new pen, since every stall has wide boards where a newborn lamb can get through. Fixing pens is not on the list. Paying for milk replacer is not in the budget. Lambing season ended two months ago, and I am finally getting caught up on sleep.

This lamb is super cute though. He's almost purring on my lap. "I'll give you a chance," I tell him, and right then that seems like the right name: Chance. His sister will be Lightning. Everyone living at or visiting the farm will be beguiled.

NEVER HAVE THERE BEEN TWO MORE SPOILED AND DEMANDING lambs. They are voracious feeders, and for the next month Chance and Lightning are utterly uninterested in anything but milk, which Kyla, Wren, or I mix up from a powder and give to them in human baby bottles eight, then six, then three times a day. They bond to humans, not sheep, and no matter how hard we try to get them to move out to the field with the flock, they will not go. Without a mother to introduce them to the marvelous salad bar that awaits in each field, they seem to have little interest in or knowledge of how to graze. They associate food with us, and we have not been getting down on all fours in the grass. They follow us everywhere on the farm, climb our legs and eat our clothes, nibble on the lettuce and taste the blueberries. Folks from all over who come to stay in our refuge village are charmed by Chance and Lightning coming over for a head-scratch—the lambs close their eyes and wag their tails in bliss, lie down and rest their chins on a human thigh.

 Peter starts taking people down in the field late in the evening as the stars are coming out, people who've never touched a sheep, or sat in long grass, or felt the grasshopper chorus fill every pore, or seen so many stars. He tells them the story of how the lambs survived as orphaned newborns way out there in the storm. He loves it. He's someone who is always moving, often tightly wound, and it slows him down. He loves to share it with others—the healing magic of sheep meditation. Most people know animals as pets—who we bend our lives around until they die—or as units of production, pushed to gain weight in order to be a bigger chunk of food. Pastoralism gives us a third way, in which the relationship with animals is so many things—companionship, love, ceremony, livelihood, food, clothing, and also an important doorway into the wild community of plants and beings that are all around us and are not so easily seen. From their words and faces, I believe that for some people, in those sheep meditations in the field, repair is happening, that something feels more coherent and more whole.

CHAPTER 7

PASSERINE

> The earth sighed a long sigh, so soft, so calm that no more than two or three eddies of birds rose.
>
> —JEAN GIONO, *THE SERPENT OF STARS*

THE BABY BIRD IN THE BARN IS FEATHERLESS, WITH wrinkled skin and an oversized mouth like a cartoon birdling, gaping wide and the color of dawn. I find it in the rafters only after searching for the smell of rot that hits my nostrils every time I pass through this aisle of the hay barn.

A dead rat, I thought, or something Rue buried in the corner for later, but I had turned up nothing in the corners. I swept the barn clean, or as clean as one can sweep a cavernous century-old barn with a heaved floor. The smell had a way of crawling on my skin and staying with me, the distinct smell of dead flesh being consumed. One knows this smell as a farmer and learns to seek it out, in case its cause is a lamb that has disappeared from the flock or a festering wound under a sheep's thick wool. Sometimes my nose leads me to save something just in time.

Not this time. By the time I think to find a ladder and search for the smell in the rafters, the baby barn swallow is long dead, flayed on the wooden beam. Something black and barely visible festers in its flesh. In the nest beside it are two other hatchlings, all skeletal.

Had their parents abandoned them for some reason, or did they have a sickness that overtook their parents' mighty efforts to feed and raise them? Had the parents themselves died or been killed? Simple starvation seemed unlikely. It is August, and this would have been the swallows' second and last brood. In another two or three weeks the swallows will leave us on their migratory jaunt to the Southern Hemisphere, but it's still not too late for the parents to find an abundance of insects in the fields to bring to the newly hatched.

The baby barn swallows in the first brood had entertained me daily for a period in June, and I had been waiting eagerly for the flight trials to begin with this second brood. The first hopeful sound was the hatchlings' crying—like the squeak of rubber sneakers on a ball court—all along the rafters under the eaves of the barn. Their squeak was the happy sound of a pivot and a jump shot, meaning a parent had arrived with a mosquito or mayfly. It meant that the blind and naked young were being fed, both parents returning to the nest with a spitball of crushed insects every ten or fifteen minutes all day long.

About three weeks after hatching, the babies' heads appeared over the rim of the mud-straw-and-spit nest as their bodies swelled to fill the space. Four or five baby birds with old-man heads fill their nest—the size of a teacup—to bursting, and sometimes one spills out and hits the floor before it has a chance to fly. If they chance to fall into the barnyard where the chickens free-range, they will be snatched up by a zealous artful dodger hen who sprints away with her prize, all the others cackling and bobbing in mad pursuit. Once, a baby swallow landed on the forgiving velvet plush of Habibi's back when he lay in the shed one morning. The birdling was still there, beady-eyed, when I herded the sheep out to pasture. It might have been my imagination, but I could swear that long-legged Habibi got carefully to his feet and walked with extra dignity and intention, like a camel carrying a tiny prince.

In my nonscientific observations throughout the years, it seems that about two thirds of the birds make it from the egg stage to flying off into the glare of day to encounter the real dangers of life as a songbird—visible in the shape of an owl or a cat and invisible in the rising tide of habitat loss and pesticide and pollution. This nest of carnage on the beam lowered my average. None of these babies had developed wings.

A bird is not born knowing how to fly. Not exactly. Leaping off a rafter and opening two perfectly constructed aerodynamic wings will get a fledgling only so far—usually to another rafter, or a spot on the ground, or sometimes to a confusing corner of a window where an invisible cobweb will wrap its sticky strands around a beating wing and mangle the delicate microzippered fibers ever so slightly so that the wing no longer beats at all.

One of my greatest joys is watching the parent barn swallows—so elegant with their ruby-colored cravat at the throat, creamy shirt coat, and long blue coattails—coach their young how to transition from being perchers and reckless lurchers to flyers capable of not only defying gravity but of snatching a transparent mayfly on the wing.

In June, the swallows had found a broken-down manure spreader in the pasture to be the ideal launchpad. As I passed by on my way down the hill to check on the flock in the morning, I would see eight or ten immature birds—like toddlers in retro pile coats with oversized heads and beady eyes—lined up and ready for school. Every time an adult swooped in with a snack, the round-bodied babes would chatter excitedly and hop sideways up and down the edge of the manure spreader. Then they would launch after the parent bird, more nervous chatter ensuing.

One landed in the grass, got stuck, and cried hysterically for a parent to come and help, though I wasn't sure how they could help, other than to fly at the head of a predator and be a distraction to avert disaster. They were swooping at *my* head as I stood there watching. A baby has to struggle back up on its own. Or not. I wondered if it would be wrong to pick the little thing up and place it back on the rim of the manure spreader, just as I had wondered with the fawn.

A few weeks earlier I had stepped over the sheep fence early one morning and heard a low snort from the woods behind me; a doe stamped her foot and didn't move away. The grass was long and dew-soaked, and my jeans clung to my thighs. As I walked to the sheep I almost stepped on a newborn fawn the size of a cat, with eyes like wet river stones. A small emerald beetle was crawling over its golden flank, but the fawn didn't flinch. I froze in midstep. Though I was inches away, the fawn's best defense was to make itself invisible. I have heard—and it seems right in some cases, but not all—never to touch a wild animal's baby, or its parent will reject it. But this newborn fawn was on the other side of a hot electric fence from its mother. Even if the mother—who can jump in and out easily—came back for it, I knew the fawn would not be able to follow her out of the enclosure.

I've watched sheep die a horrible death in electric fence before I could cut them free. I once released a badly damaged wild turkey, and I've found the fried carcasses of spotted salamanders who crawled over the bottom wire in their spring migration. So I reached down and picked up the fawn—who weighed less than a newborn lamb and was limp and warm, with a dappled coat—and I carried it to the edge where I had seen the doe in the trees. She was gone, but I left the fawn in the ferns. Later, I saw them walking together under the apples.

When we came to the farm in 2001, the air was thick with swallows. Ann Day walked me around the outside of the barn and showed me how the barn swallows took up residence each May in nests along the southern and eastern eaves of the large roof, while the cliff swallows made their nests along the western side. A few tree swallows claimed the north-facing entrance to the barn under a long covered ramp extending from the eaves, but mostly they occupied the nest boxes Ann had put up all through the fields.

Ann, who was an expert birder and for decades had written a weekly nature column for the local paper, taught me how to identify each type of swallow: The barn swallows can be easily noted in flight for their striking forked tails and are indigo blue with a creamy underside and russet throat. Tree swallows are elegantly simple: blue-black with a white breast. The cliff swallows are dressed in soft, earthy tones like the barn swallows but have a less rakish tail—notched instead of forked—and a small white forehead patch. Their nests are different too. While the barn and cliff swallows build theirs out of mud and straw, the tree swallows gather twigs, grass, and bits of bark to inhabit the boxes in the fields. The cliff swallow nests are nearly conical ovens of clay with a little hole in the side facing the sun, while the barn swallows make more open bowls attached to ceiling beams.

The year after we moved in, I counted sixty-three nesting pairs of swallows and noted it in a journal. In recent years I've counted a fraction of that number, even as low as eight. I learned that all types of swallows are in decline in North America, as is the case for all avian aerial insectivores (birds that catch insects on the wing). Major reasons, according to the Audubon Society, are fewer insects due to pesticide use, and a loss of nesting sites because of the loss of old farms. In one study I came across, a significant factor in the swallows' decline seemed to be the rate at which hatchlings died before leaving the nest. But researchers didn't pinpoint why. This is what I wanted to know. Why had all these baby birds died?

I go up the ladder, pick up the dead bird on the rafter gingerly by its feet, and place it in the nest with its mates. Then I dislodge the nest from the beam. I stick the mess under my arm and carry it outside below the barn, where I tip out and fold the birds into the soft darkness of the giant compost heap. Kicking a pile of warm, half-decomposed hay and manure over them, I am comforted to think that whatever ailed them will be cleansed by the death-eating microphages of the living soil—dirt as purification. I crumble the nest onto the compost, a salad of dung and dust, bone and beetle, bits of pearly shell, and shards of grass. Each material has its own life history, its own constellation of elements and matter essential in the creation of life on earth as well as in the twinned cycle of decay. These elements came together in these forms—egg, bone, feather, flesh, heart—in this moment, to be taken apart again and return in other forms, without limit into eternity.

SWALLOWS MATE FOR LIFE, AND EACH YEAR THEY ARE thought to return to the same place for summer breeding and raising their young. Tree cavity, barn loft, scent of hay and moss and sap are retained in somatic memory as a homing device. Those that are born at our farm, then, should return with their parents, increasing the community, and yet fewer come back each year. I find it fascinating to read about how birds navigate by celestial bodies or magnetic fields or how salmon find the stream of their birth by smell, but the science of it all matters less to me than the way I recognize this instinct and believe it must be real; blindfold me and dull all my senses and bring me to a hundred farms in the dead of night, and I would still be able to sense which one was home. This story of being hefted to a single place—which may be as much a cellular intertwining as a conscious making of meaning—is something the swallows and I have in common.

As I stand on the compost pile, a pair of barn swallows chattering and swooping low above me, I ponder the evolutionary advantage of returning to the same nesting ground—even the same nest—year after year to bring forth the next generation. A place-based literacy, an intimate knowledge of that landscape and its moods, its food and shelter, would be beneficial to a species, surely. But it seems odd to make a dangerous and ravaging journey over thousands of miles if other places along the way might be just as good to bunk down and procreate. Is what we call attachment any different than this micro-adaption to place? Though we define attachment as something felt and formed over time, something emotional or spiritual, surely the deepest forms of attachment happen on a cellular level, in people as well as in birds and fish. To recognize a place in your bones, marrow deep, to feel beyond knowing.

THERE ARE TWO WAYS THAT I MEASURE DIMINISHMENT IN the natural world, a world we all have the ability to see and sense, no matter where we live. The first is ecological: a loss of vitality, complexity, and stability. This can be studied and measured, but it can also be perceived by simply listening and noticing. Nature has a voice that sings in different registers, and in those registers you can hear health or struggle, presence or absence. As a farmer, or anyone else who spends time daily on the same piece of ground, you can sense regeneration or decline.

The second way I measure diminishment is in the human experience: loss of beauty, of meaning, of pattern language. These also become more available to us as we watch and listen, take in what surrounds us. These are found in the words and languages of people who live by the land. "A lexicon of land-love," to use Robert Macfarlane's lovely phrase.

Last summer we built a wooden stage near the garden to hold concerts. The first performers came for a week. Our gift to them was food and a beautiful place to stay and practice; their gift to us in exchange was music. The night before the public show, they invited Peter and me out to the stage to hear a new song. Kat, the lead singer, wore a flowered sundress and bright-red lipstick, her black hair sweeping across her forehead and flowing to her waist. She had a voice as soothing as honey, as precise as clove. She sang: *Well it hurts to be a butterfly, baby / you got to step into your cocoon and let it all go / fold your hands and let it all turn to liquid / you got to let go your fears if you want to fly.*

Peter looked up and saw a monarch chrysalis hanging from the newly milled roof beam of the pavilion that he and Ky—our neighbor who built the frame—and my nephew Chase had just put up two days before. The chrysalis trembled and shone like a raindrop on a green leaf, hanging by a thread so slender that the tiny, gold-ornamented crypt, inside which a caterpillar was being liquefied and remade as a butterfly, seemed to hover in the air.

The chrysalis survived the concert, and people marveled at the coincidence: Had Kat's song summoned the caterpillar to do its transformation there? It made the metaphor that much more powerful. And yet no one noticed when the butterfly emerged the next day. She fell fifteen feet onto the barren wooden platform, crumpling a wing. I found her the next evening, nearly lifeless, and carried her to an Echinacea flower to take her first sip of nectar and warm her wings in the last of the day's sun. Nature, I thought, is increasingly a metaphor for our own troubles, which is powerful medicine for us, but it doesn't help nature. It's logical that as nature dies around the world, we engage with her more often as an abstraction, keeping her alive in our imaginations only. But this won't be enough.

TAXONOMICALLY, SWALLOWS BELONG TO A LARGE FAMILY OF songbirds called Hirundinidae, which includes martins and saw-wings, and to the order Passeriformes, called in the vernacular passerines. In the Aves family tree, the passerines are one of the main lineages of perching birds, so classified for the shape of their toes. The word *passerine* itself, however, conjures the image of passing through or above, making passage, which I love for all the images I have of them swooping through doorways and openings or passing above me in front of a storm cloud, chasing insects on the wind. Because swallows return home from the far reaches of the globe, English sailors tattooed their chests and arms with images of swallows as talismans for safe passage home from a dangerous voyage at sea. Swallows being a bird of the land and farming country, they were a welcome sign to sailors that the ship was close to port after a long ocean crossing.

It was once thought by naturalists that swallows never migrated at all but instead disappeared for winter only because they buried themselves in mud or caves and were then resurrected from the earth itself. This is a belief passed down from Aristotle to Linnaeus to Gilbert White, who held onto this idea until the mid-1700s—"It is much more probable that a bird should retire to its hybernaculum" than migrate, he argued in his classic book, *The Natural History of Selborne*—even as other naturalists debunked it. Centuries before, Swedish writer Olaus Magnus (1490-1557) wrote, "As summer weareth out, they clap mouth to mouth, wing to wing, and leg to leg, and so after a sweet singing fall down into great lakes or pools amongst the caves," adding how ice fisherman would find them in frozen clods along the banks and take them home to warm them in their stoves to restore them to life.

Was this belief perpetuated out of confusion? Swallows were very numerous back then, and perhaps some who were caught in an early cold snap before the migration would have been found in a kind of torpor, huddled for warmth out of the wind. Or did this belief stem from a human need to transfer a spiritual longing for transformation out of the darkness onto a creature already thought to be mystical because of its aerial life, close to the heavens? Probably both. Winter in the Northern Hemisphere was cold, long, and most of all, dark. People would have missed dearly the infinitely various sounds and daily flight patterns of songbirds and longed for them. Even one story of a fisherman bringing home a frozen bird and reviving it by the fire would have become a tale mythic enough to sustain a whole village through the winter; it would have reminded them that summer's sweeter days weren't entirely lost, only resting beneath the ice, mystic chords of memory waiting to be found, like our better angels.

Maybe our instinct to save wild creatures is most of all an effort to save ourselves: an attempt to summon and restore the hope and altruism that gets us through times of hardship. To feel a tiny flutter of life in one's palms after coming in from the cold is the greatest hymn one can hope for.

A FEW WEEKS AFTER FINDING THE DEAD BABY BIRDS, I mentioned them to my friend Anna, who suggested they might have died of mites, tiny parasites that live on a bird's skin and suck away its blood and moisture. It was early September by then, and I thought all the swallows had gone south. Though it still felt like full summer to us, I imagined that swallows could gauge the drop in hatching insects better than the best fly fisherman, and food for aerial insectivores in the North Country was beginning to get scarce. The barn was empty of the swallows' chatter and squeaking, their fluttering of dusky wings in the rafters, their fresh white streaks of guano on the old boards. This, to me, is the saddest time of year.

I decide to clean out the tree swallow nest boxes in the fields so that mice and mites don't overwinter there, and the birds can build fresh nests in the spring. The boxes were made with pine years ago and are in advanced stages of decay. Each box has a wooden front panel with a small round hole cut large enough for a swallow or bluebird but too small for a blue jay or starling—birds that might rob a nest of its eggs—and this panel lifts up and down on a top hinge. It locks shut at the bottom by a bent nail that you turn up.

The first nest box I come to has come loose at the bottom, and twigs and other nesting material bulge out. It is easy to lift the front panel and scoop out the foot-wide mass of several years of nest building, which I hold in my hands to admire. Coarse twigs are piled with finer ones, then soft coppery pine needles, then delicate strips of birch bark as fine as ribbon. A rim of fine sheep's wool. A piece of string from a grain bag. At the heart of the nest is a deep swirling galaxy of black feathers.

I look at them more closely, marveling at the bird's instinct to collect only black chicken feathers to make a camouflaged hiding place, when suddenly I see those loose black feathers arrange themselves into the shiny midnight-blue back of a swallow—the color and sheen of a lake at midnight reflecting the stars. Absolutely still. Black feathers in the shape of a perfect arrowhead. Something you don't see until you see it, and it pops out amid the shards.

Is she dead? She doesn't move, so I tentatively and ever so lightly place one finger on her jeweled back. It is warm. In another split second she is gone, straight into the sun.

Exposed in the swirl of midnight feathers are four pearly white eggs. I feel an overwhelming sense of grief. They look up at me, innocently, and I look back at them with the eyes of death.

Once a family friend brought Wren a tiny darling nest woven with pine needles and birch bark with four tiny unhatched eggs, cinnamon speckled. Though she was very young, it fit in her hand. She held it very still, didn't say a word. "They are wren's eggs," the friend said. "Like you."

When the guests were gone, Wren told me it was a sad nest and she didn't want to keep it. "Let's put it outside," she said.

I think of that as I put the swallow's nest and its eggs back into the box just as it was, and repair the door. I say an apology, look into the sky, and pray the bird will return to hatch her late brood. Then I turn to go back to my work pulling out bindweed in the orchard.

As I walk toward the blueberry field, staring down, I see a slender snake moving strangely in the emerald grass. We see a lot of milk snakes and garter snakes in the orchard; to me they are always a sign of health in the land. I crouch down and see that this young milk snake—about the size of a shoelace—has completely tied itself around a sturdy stem of timothy. Its body has tied itself in a knot, around itself and around the grass, in a wad of green sinew and gold skin. The more the snake thrashes the tighter the shackles of grass become, cutting into its delicate scales. I kneel down, cup one hand over the snake to calm it, then break the wiry stem with a quick snap of my forefinger and thumb. The snake slithers from a snarl to a sinuous line, back to its mysterious, shifting galaxy of grass, back to its miraculous, more-than-metaphorical life.

INTERLUDE II

The sky was smooth as a washing stone.

—JEAN GIONO, *THE SERPENT OF STARS*

THE SUMMER AFTER I FIND THE DEAD SWALLOWS, WREN IS fifteen, and she learns to drive stick shift delivering blueberries in our old farm truck. Every Wednesday we break away from our picking and other chores to take flats of blueberries into the valley to stores and restaurants that order our wholesale fruit. We deliver to American Flatbread, the Pitcher Inn, the Warren Store, the East Warren Market, and Tom, at the Featherbed Inn, who makes pies.

Wren drives the crooked roads, her eyes glued straight in front, her back not touching the seat. While she is getting used to the clutch and shifting gears, the truck does not so much roll as hop down the road like a sheep on a rope, sometimes out of nowhere stalling to a dead stop. Over my shoulder, I nervously watch the tower of cardboard boxes, each filled with pints and quarts of berries, seat-buckled into the narrow back seat of the truck cab.

Given that we live in the mountains, with no traffic lights but plenty of stop signs at the bottom or at the top of steep hills, learning to drive a manual is genuinely terrifying. Right away Wren got good at rolling to a smooth stop and not stalling at the end of our precipitous road, about a half mile downhill from the farm. Starting up again while tilted at an angle at the stop sign, and doing it quickly to fit into traffic on the main road, is another trick.

We wait for a while at the stop sign, even though no cars are in sight. I don't say a word. Wren revs the engine to a roar, bravely jams it into first, lifts the clutch and simultaneously pushes down on the gas, and we leap forward like a horse out of the starting gate, then shudder to a crawl just as suddenly until she finds second gear. My eyes are glued on the back window in case someone rams us from the rear end.

"You're driving a little erratically. A little slower on the clutch."

"I think I'm driving really well today."

"I'm afraid these berries might tip," I say, turning around in my seat with one hand on the boxes brimming with blueberries, and Wren points out that she could really use some help on traffic rules.

"Don't hit any old ladies," I tell her as we drive into Warren, usually full of white-haired tourists on beautiful afternoons.

Once we get all the berries out of the hot car and safely to their destinations, I suggest we turn south.

"It's so damn hot. Let's go for a swim."

"Don't you have to get back?" Wren asks.

"Yeah, but you need driving practice."

"More is always good."

DRIVING TOGETHER THROUGH HILLS THAT ARE FLEECED WITH humidity and shimmering with heat, exhausted as I am from so many straight days of harvesting in the sun, I feel a surge of exhilaration to be wandering together off the farm. All the days have run together, weekday and weekend a nearly indistinguishable whirl of people coming to the farm to pick berries or stay for a program, staff who live with us needing direction and to-do lists, cooks making meals in the house and groups gathering to eat them in the top of the barn. Even the sheep have been almost an afterthought in the stew of summer harvest season.

Too soon Wren will go back to school. When Willow moved away to college in the Midwest last year, Wren decided she needed an adventure of her own. We had raised a child who was self-assured and physically fearless, could solve problems with her hands as well as her mind, and could have deep conversations with anyone from any walk of life, but whose learning was entirely unconventional and undisciplined. She returned to public school at twelve to have more time with friends, but she was frustrated by so many long hours indoors without much to show for it. She wanted "something rigorous," she said, so to our surprise she sought out a private school in Massachusetts. Her first year had been desperately hard; she was full of homesickness and overwhelmed at being so far behind. Most of all, she wasn't in sync with her media-centered generation. So many of the things she loved most were of no consequence to anyone. She couldn't find people to mess around in the woods or swim in the river in the rain, and she felt terribly lonely. The place felt confining, demanding, and too far away to come home for weekends. More than once, Peter and I had driven three hours to have dinner with her and cheer her up, then driven three hours home.

With summer nearly over, I'm filled with the sense of needing more time with her, and she with me. Even though she worked on the farm, I was busy from dawn to dark and never took a day off. Except for these driving lessons, I barely left the hillside. An image: Wren at four, crying and saying, "I know you're here, but I can never find you."

That was just yesterday it seems. How has the river carried me so far, so fast? Soon my baby will have her license. I want desperately to stop time.

Barry Lopez was once asked to give advice to a fifteen-year-old girl who wanted to be a writer. His answer: "Read a lot. Find out what you truly believe in. Get out of town, leave the familiar." *That seems like good advice for every parent to give every teenager*, I thought when I read his words, but it also shatters me. Like me at her age—and much longer, until I was well into my twenties—Wren struggles to leave home. A huge emptiness wells up. A sense of pure panic. I have heard it in her voice over the phone. I have felt it in my own chest, and I know I must help her to overcome it, even as I have also helped her to attach to this piece of ground. She will find strength in knowing that she can leave, that she's capable, and that she can return. Me, I'm trying to remember that letting go of something is not the same as losing it.

WREN CUES UP OUR FAVORITE SONG ON THE TRUCK AUDIO, and I crank it up. *No one on my shoulder bringing me fears. Got no clouds up above me bringing me tears.* Patty Griffin's strong but vulnerable voice fills the car. *Got nothing to tell you, I've got nothing much to say. Only I'm glad to be with you on this heavenly, heavenly, heavenly, heavenly day.*

We follow the winding road south out of Warren and drive through a heavily forested steep section called the Granville Gulf, hot wind buffeting through the windows. The road narrows, and all along it are rock ledges covered in moss and streaked with water that comes from dark shaded hills high above. The air cools; even in summer there's the smell of snow.

INTERLUDE II

We turn off the pavement onto a dirt forest-service road. A creek tumbles down through a steep valley and over sculpted granite, the water glinting green and silver below the roots of trees. We pull over, kick off our shoes, and scramble down the vertical shelf from the road to the gorge, hanging onto maple saplings and digging our bare heels into the dark forest duff. We peel off our clothes and slip our tired, berry-stained, sweat-stung bodies into the achingly cold water in the broken forest light. The river flows over us and beyond, holding us afloat, this moment. *This*: a word like an outstretched hand.

CHAPTER 8

SHEARING DAY

You didn't see them, you heard the noise of their cascade, and the shepherds' whistles, and the swaying of the lanterns.

—JEAN GIONO, *THE SERPENT OF STARS*

SHEARING DAY BEGINS IN GLOOM, THE WIND FULL OF spitting rain. I flip on all the lights in the barn, though it makes little difference to dispel the dark. Hay fills all the space in the center bays, right up to the rafters with their quiet nests, blocking the light and muffling sound. Summer is dried and baled now, to come again one day. Layers of cobwebs gauze the window wells, heavy with dust.

I pull down an extension cord from the wall and set up a work light on the beam above the shearing board. Then I sweep out the hay chaff and piles of dirt that come up from the cracked cement floor, stack the odd assortment of water buckets and broken chicken feeders against the wall. I find two sawhorses and a tabletop made of wire. This is our fancy skirting table where we will hand-sort and pick through each fleece for burrs and dried dung before we send it to the spinnery for cleaning and making into yarn.

A huge pile of unskirted fleece from last spring's clip still fills the fleece bay. We shear twice a year, in May and October. The spring wool is always troublesome; because the sheep have been eating at the hay feeders all winter, their wool is full of fine chaff that is impossible to remove.

Their wool has often felted, too, like a dense rug, from living all winter in the snow. To make it useful for any type of yarn would be so much work that we'd lose gobs of money doing it. Even the community wool pool where we can drop off fleeces to make a few cents a pound doesn't want such ratty stuff. We use it for mulch, mostly. We've given spring wool to friends to insulate a sauna, and once to a young artist, Isolde, who stuffed it into the walls of her VW bus to provide substrate for growing mushrooms. We got to be friends with Isolde, and she parked her bus and lived here for a while (with her friend Brandon, the palest blond I've ever seen, who would streak like a comet across the lawn into the pond. "Does he have to?" Peter would ask). Through a tiny "truth window" in the inside wall of their bus, I saw the white strands of mycelium growing over the wool. I never understood the reasoning behind it.

I go inside to grab another layer. Maybe a hat. We'll warm up soon enough and strip it all off, but for now I have a chill. I put the kettle on, make another cup of PG Tips. I pour some water and oats into a small saucepan and put it on the flame. Then I sprinkle in some cinnamon and a pinch of salt, slice up some apple, plop in a handful of blueberries from the freezer, and give it all a vigorous stir. The creamy oats turn purple and start to erupt, blowholes letting off little jets of steam like something at Yellowstone. Peter doesn't like to even look at oatmeal, but I eat it nearly every morning, with creamy yogurt and maple syrup poured over the top.

Standing at the stove, I see Zoe come around the corner of the house, heading for the barn. Zoe has a climber's grace and a bull rider's toughness, wild red hair and a huge laugh. Run into Zoe, and you will instantly feel better about everything. But today she is holding her arm, looking less ebullient than usual.

"I'm thinking of taping my arm to my side," she says, coming in the side door. "Maybe I can help one-handed."

"What happened?"

"Oh, just being a dope. I was doing strengthening exercises because I feel like my shoulder is about to pop out, but I think I overdid it."

Zoe grew up a few towns away and became part of the farm when she was dating a young friend who lived with us one spring. They broke up and Zoe stayed. She lives in a tiny house she built across town, but she moved her tattoo and art studio into the basement of our small barn, where she makes beautiful drawings on everything from concert posters to snowboards to both small and vast portions of people's bodies. When she has time, she loves to help me—the more physical the job the better. Usually, that is.

"I can get the duct tape," I say. "But we can also try to go easy on you today."

I am eating my oatmeal standing at the sink. I get through half a bowl, and the anticipation of the day suddenly fills me too much to continue, so I throw the bowl, spoon, and pot with the rest of the now-congealed purple gloop into the sink to deal with later (sorry, Peter), and we head back out the door.

A DAMP SHEEP MAKES FOR SLIPPERY HANDLING AND CAN TURN a difficult job into a dangerous one. I promised my shearer, Gwen, that the sheep would be penned all night out of the rain and mist. She told me she had just finished shearing hundreds of sheep at a big farm in upstate New York and felt "stiff." She said she didn't want to have to wrestle a wet sheep thrashing on the boards, because she could easily pull a muscle and hurt herself.

"I'm trying to look out for my body in my old age," she said.

I have never heard her say anything so radical.

The flock are resting in the shed, and Zoe and I walk among them, feeling into their coats to make sure they are good and dry. We crowd them up into a smaller section of the shed so that we can grab each one for its haircut without chasing them, getting them excited, or putting extra strain on our backs. A sheep who doesn't want to walk is unbelievably strong and hard to move forward.

I'm glad the flock gorged on lush grass in the orchard yesterday, because today they will get very little to eat. Sheep whose bellies are nearly empty will feel less uncomfortable and handle better. And poop less. Robert Livingston warned in his 1809 *Essay on Sheep* not to tie their legs together, for it "forces the sheep into a position with the intestines being pressed . . . they discharge their urine and dung at the time they are sheared, which fouls the wool and is offensive to the operator."

"Just don't feed them," Gwen, my operator, always says. But I sneak them a little hay for this special day. I love fall shearing. The spring shearing is about keeping the animals healthy in the heat, but fall shearing is our harvest, and it feels like harvesting light.

This is the wool we handle with care and take to David's spinnery to be turned into lofty single-ply yarn in gorgeous shades of silver, nut brown, brown black, and cream. Icelandic wool isn't the softest; it's a dual-coated wool that evolved in cold climates, with a soft undercoat and longer, coarser guard hair that sheds the rain. If you can separate the tog (outer guard hair) from the þel (pronounced *thel*) and just spin the softer, shorter þel, you can get a beautiful and extremely warm yarn. Either way, it has what fiber people call "character" and some call "tooth." It has a luster and a halo. If you were a chef, it would be like cooking with something that was foraged from the wild—it comes with a lot of variation and attitude, and it's totally worth the trouble.

THE SUMMER I WAS AN UNCERTAIN WAIF OF THIRTEEN, MY mother and I made a woolen blanket together. It had plenty of tooth. It was a project, I imagine, meant to distract me from myself and my pointless bouts of melancholy and was a way for her to practice her newfound obsession with spinning, under the tutelage of our friend and neighbor Donella Meadows. Dana, as we called her, kept a small flock of sheep that I barely remember, eclipsed as they were by her terrifying geese, who came after me with their necks outstretched, hissing like serpents.

Dana was a brilliant systems thinker—she coauthored a best-selling book, *The Limits to Growth*, in 1972, predicting the crash of the earth's systems—but she was a peasant at heart. She loved more than anything to make baskets, and to spin, and to grow things with which she could make baskets and spin. She, my mother, my sister, and I wove fences out of saplings, made baskets out of willow and reeds, and even tried to grow flax to make linen (a complete failure).

Dana and my mother made an Ashford wheel from a kit. I still have the wheel, which is made of pine, leather, and string, following a design from the early 1800s, when every female farmer in rural America was making her family's clothes from wool she spun herself, later to knit or weave. You sit on a chair with the wheel in front of you, your foot on a wooden pedal at its base. You start the wheel spinning with a shove of the hand and keep it going with your foot, tapping out a slow steady rhythm. A string around the big wheel loops around a small wheel that holds a spindle. This is where you attach a piece of carded sheep's fleece—or roving—and pinch your fingers along it as it spins to twist it into a fine thread. If you go too slow, the wheel starts turning backward; too fast, and it will suck and break your thread out of your hand.

It took me all summer and many hot tears of fury to learn how to make a lumpy and useless spool of yarn. It felt a lot like trying to rub circles on my stomach and pat my head at the same time, something I was never good at. But I liked the challenge and the rich smell of lanolin from the fleece on my fingers, and most of all I wanted to participate in something my mother loved. The sound of her spinning wheel, with the steady beat of the wooden pedal on the floor making a bass tone to the higher whirring of the wheel, was as soothing to me as honey on a spoon.

Before I learned on the spinning wheel I learned the oldest form of thread-making of all, on a hand spindle. Some version of this smooth stick with a disk on the end is found in pastoral cultures all over the world. The first cloth was most likely made by people living in the mountains, rooing fleece from where it snagged on branches and rocks from wild sheep, twisting it with the weight of the spinning disk and winding it around and around the shaft. I had a very simple spindle made by a woodworker at our local fair. I had no idea that I was using an ancient technology that, in the hands of Indigenous women, also became a source of spiritual power. Symbols of salmon, bear, and eagle carved into the round disk of the spindle by tribes such as the Coast Salish of the Pacific Northwest were carefully chosen to spin that animal's power and personality into the item of clothing or blanket that was being made.

My mother spun most of the wool for our blanket, unwashed, after it was shorn from the backs of Dana's Dorset sheep and combed into roving on hand carders. My job was to research and collect most of the plants to dye the skeins she made, from the fields and forests around our farm. We had a book from the library that named many plants I recognized in the fields and others in the garden, but to turn them into dye was something I'd never thought about before.

I'd walk around with a burlap sack, stuff it full of flower heads and leaves, even roots and lichen and the shells of butternut, then come back to the kitchen, where my mother and I would shred and cook the plants in giant pots, filling the house with strange astringent odors that lingered for days.

Dana grew tansy, a button-shaped yellow flower that had been used since colonial times to make muslin brighter. Goldenrod collected in late summer made a rich orange or a pale ochre, depending on which metallic salt—or mordant—we fixed it with in the dye bath. Onion skins were a deeper yellow. From one plant we got a beautiful sage green, but I don't remember now which plant it was. My favorite dye was a rosy gray from an unidentified lichen I scraped off a rock ledge in the woods. All of the colors were subtle, and some variation of gray green or orange yellow. It seemed to be nature's palette. I loved the subtlety myself, but maybe my mother needed more excitement, because somewhere she got her hands on some indigo, made from the fermented leaves of the indigo plant, and some cochineal, a soft-shelled beetle that feeds on prickly pear and turns things red.

Once she had knitted all the skeins of homespun dyed yarn into six-by-eight-inch rectangles with moss stitch, my mother and I stitched them together. The four squares of blue indigo and two of red cochineal stand out on my finished blanket like vagrant tropical birds in an undulating field of pale green-gold wheat.

My blanket was one of the only things I took with me when I went to college. The smell of it filled me with nostalgia, but most of all it awakened in me an enduring reverence for making things by hand. I still have the blanket, forty years later, full of open seams and mothy holes, but the colors remain as I remember them from childhood.

"You need to darn that darn thing, Mom," Wren says if I pull it out of the chest where I try to protect it from further disintegration. Only to me is it a thing of enduring beauty.

I CAN TELL FROM THE DOGS' EXCITED BARKING OUTSIDE THE barn that Gwen has arrived. Her small black shepherd, Lyle, is a cousin of our new puppy, Machias. They both come from Elaine Clark's sheep farm in Maine and share a genetic quirk of flashing a side-tooth grin and waggling side to side when feeling delighted. I bartered a hundred pounds of wool (for Elaine's rug-weaving business) for eight-week-old Machias—a little pointy-faced, golden-eyed arctic fox of a dog with a ridiculous puff of white tail and a bad case of FOMO. His barking drives Peter mad.

The pups are overcome with excited sniffing for the first five minutes and then start strutting to impress each other, kicking out their back feet and sending scraps of sod flying. It's a comical display of boyhood. Rue, with substantially more dignity and girth at age eleven, stays resolutely at Gwen's truck door, looking for cookies.

Gwen is grabbing the tools of her trade out of the back—clippers, the engine that runs the clippers, a metal panel that screws to the wall so she can shear anywhere that an extension cord can reach, a tool kit with extra combs and oil, soft suede shearing shoes that allow her to slide on the shearing board and to feel the sheep through her feet.

Gwen is tall and very lean, dressed in black jeans, two layers of black shirts, and a blue cotton headband wrapped a few times around her long fine brown hair. She has the lithe, taut body of a dancer and the huge, uncoddled hands of a blacksmith—knobby fingers, black nails, broad palms, and splayed thumbs betraying her tens of thousands of hours of handling sheep on the boards.

Today her left hand is red and swollen, the skin tight over her thumb joint and wrist.

"What happened?"

"I don't know. I got kicked, I think," she says. "It's all right though."

I get it. I, too, often get banged up and forget, by the end of the day, how it happened.

Among the four of us—Zoe, Gwen, and Gwen's partner, John, who has a serious case of Lyme disease that has caused his knees to swell—I am the only one without an injury. Yet. The day has barely begun.

We all joke about our sorry physical state for a while.

"Your sheep look big," Gwen says as she looks through the barn door.

"Oo, and that one over there looks mean. She's going to beat you up," John says. John is an Irishman with a perpetually skeptical expression who doesn't say much at first, sort of hangs around the edge of a conversation and then slays everyone with a deadly dry joke delivered in a soft lilt.

"C'mon," I say. "You're psyching yourself out today."

GWEN LIVES CLOSE TO WHERE SHE GREW UP, ON A HOMESTEAD in New Hampshire. In the early years of her childhood the family lived in a half-built cabin without running water, taking bucket showers on the porch even in winter. Her father, David, sheared—a business he had begun passionately at the age of fifteen—and also taught school, while her mother raised three children. David, who people still talk about because his stories and wit were legendary, always invited Gwen and her brothers along on his shearing road trips.

"My older brother sheared exactly twice and still has zero interest in farming," Gwen told me. "My younger brother would shear now and then. I came to it last, reluctantly at age twenty-five, but with me it stuck."

Shearing is one of those trades you learn by apprenticeship. People say you need to shear at least 1,000 sheep before you have a clue. But there are schools for it, and places—like New Zealand—where you can hone your skills in a trial by fire. Gwen worked there on a crew of thirteen roving shearers for a season, on farms that are huge by American standards: 16,000 sheep is a respectable flock. Shearing with a contract crew is a competitive sport, with shearers racing to finish the most sheep. The pay, fifteen years ago, was seventy-five cents a head, Gwen said. Being the only woman, and at age thirty much older than the guys, Gwen's nickname was Grandma.

"But I kept up by the end. My best day was shearing 350 sheep by quitting time . . . which almost killed me."

John is a blade shearer, which means that he gets the wool off the sheep the old way, without electricity, with a set of shears that look like giant scissors. He is so good at this that he just got back from the Golden Shears international competition in Scotland—the Olympics of sheep shearing—having qualified to be one of two blade shearers to represent the United States.

"How was it?" I ask, as Gwen sets up to shear. I have never been to a competition, but I can get excited about all the stats, marveling at how people can shear a sheep in under a minute, then do it again and again for hours. Any kind of obsession fascinates me, and shearing particularly so because it's a professional job that's been made into a competitive sport, rather than the other way around.

"With the even number of blade and electric shearers, and over thirty countries competing, it made blade shearing seem almost normal," says Gwen, who was there with John.

John makes a face behind her back. Blade shearers are used to being made fun of for carrying on such an antiquated tradition.

OTHER THAN THE CHANGE FROM HAND SHEARS TO ELECTRIC ones (which are still hand-held, but faster) in the mid-twentieth century, very little has changed in the way we harvest wool. No matter how large the flock, we still need people to handle and shear each sheep individually. In *Très Riches Heures du Duc de Berry*, an illuminated book of hours made circa 1412, a painting shows two shepherds sitting in a field with a flock of sheep, taking off their wool with a hand tool absolutely identical in design to what John uses today.

The problem of how to get a clean wool clip off their sheep and make a respectable cloth enormously vexed the colonists of the New World, even though most of them came to America from Europe—a place where wool had been the dominant driver of the economy since medieval times, responsible for the vast fortunes of the most prominent merchant families and constituting a sovereign trading source for kings and churches for centuries. The situation in the colonies could not have been more different.

Not only did the colonists fail to raise sheep that produced quantities of good, soft wool, they did not have the infrastructure to sort, wash, pick, card, spin, and weave the stuff into cloth. The first woolen factory—in Hartford, Connecticut—lasted from 1788 to 1795, long enough to make George Washington thirteen and a half yards of brown broadcloth to have a domestic suit tailored for his presidential inauguration as a political stunt. But the factory closed soon thereafter because its product was of inconsistent quality and overpriced. In 1809, the editor of the *Raleigh Register* wrote, "We anxiously look forward to the day, when a man may furnish himself with a good Coat, for either winter or summer, without being obliged to send 3000 miles for the Cloth."

Farmers of the time washed their sheep in ponds or rivers before shearing them, jumping in to rub the fleece clean with lye soap and rocks while it was still on the animal's back. They would dry the flock—hopefully in a clean place—then have the challenge of hobbling each animal to keep it still while it was being shorn. Robert Livingston, in his seminal *Essay on Sheep*, did not approve of tying their legs together but contemplated tying them to a table "after laying them on one side; but this I think would subject them to some risk if they struggle, and at all events will require twice tying, as the sheep must be turned." He then labors on for another page describing various hobbling techniques that would allow the animal to stand but not move its legs. In the end, at least he was humane: "Let the master then show no impatience if he would have his work well done. Great care must be taken not to wound the sheep." Still, it's hard to imagine the days and weeks it would have taken to get through the very first step of producing wool for the mill in colonial America.

The brief Merino sheep fever of the nineteenth century aside, when Vermont alone produced several millions of pounds of wool a year and New England was full of water-powered woolen mills, the United States has never been a producer of fine wools or woolen cloth. Today there are people who are ingeniously rebuilding small local supply chains to make local clothing, encouraging people to know their "fibershed" much as they might know where their food or water comes from. This is a heroic effort in the age of synthetic clothing mass-produced cheaply overseas, and it's slow going. We lost the infrastructure we had to make cloth many decades ago. As Clara Parkes concludes in *Vanishing Fleece*, "Whether it's shepherding or shearing or scouring or spinning or dyeing, I keep coming back to the fact that each of these links in our chain is in peril."

With the exception of a small number of large ranches in the West that have thousands of head, most of us raise sheep for the same reason as the subsistence farmers who lived here two hundred years ago: for their manure and grazing to improve the land, for meat to feed ourselves and sell in the local supermarkets, and for wool as a side hustle. This is okay. When you farm in America, there's always this pressure to scale up. Our culture doesn't respect the small. But even small hill farms, given enough of them, can make a substantial contribution to the self-reliance and local economy of a region, both through what they produce and also what they need in the way of supporting local services—vets, tractor supply stores, feed stores, seed, labor. Small farms everywhere are endangered. They are a species people don't think about until they are gone, and suddenly the place they live has less of a story, feels tangibly less vibrant, healthy, and enlivening to the spirit. Small farms are worthy and need help to hang on.

WATCHING A GOOD SHEARER, FOR ME, IS A LOT LIKE WATCHing a fire: I can't take my eyes off it. No matter how many years I've seen Gwen work, I still like to stand nearby and watch her follow the precise, fluid, and certain dance steps with each sheep. It's astonishing and beautiful to see a very woolly, round sheep rapidly transformed into a sleek, velvety form.

She's working on a white ewe lamb who came in looking like a dun-colored mop. Gwen starts by spinning the lamb in a spiral to sit her down on her haunches. She leans the animal back against her legs with its head relaxed and dropped, front legs dangling and back legs poking stiffly upward. The first strokes are to remove the coarse belly wool, then Gwen pushes the clippers against the skin up under the thick coat to the neckline to burst the wool open at the neck. Now it can come off in a single flat blanket. She works around the face and ears and up the foreleg, then gently turns the lamb on its side and steps one leg over it to pinch-hold it in a lying-down position between her feet. The ewe kicks a little but can't get purchase on the floor and relaxes again as Gwen responds ever so subtly to reposition her.

With one hand she holds the lamb's head, and with the other hand works the shearing comb in long blows up the animal's side. Gwen's legs are straight, her feet in soft leather slippers tucked under the sheep's body to hold it still, her back bent over in a deep bend while she swivels her shoulders and arms in long sweeps along the sheep's body lying on the board. Next she sits the sheep up again and works her way along the spine, up the second front leg, and down the other side, finishing with the hind flank. All along, one hand is pulling taut the skin so that the blades can run smoothly and quickly along the body without making a nick.

Though it looks fluid and elegant, everything about this handling of the animal requires tension, strength, and endurance. This method of carefully scripted steps was developed in New Zealand in the 1950s and was readily adopted by shearers around the world for its speed, the safety it gives to the animal and the even quality of the resulting fleece, which falls off in one unbroken blanket.

The lamb's fleece grows on the board with each long stroke, revealing the bright-white underside that has never seen the sun. Soon the naked lamb is lying in a bed of her own luxurious material, as if having shed a satin gown. The fleece shimmers with the light of all the spring and summer mornings of dewy grass from which it came. Wool is an alchemy of animal and plant, and a miraculous one—a lofty fiber that is warm when wet, resistant to rot and mold, and strong enough to make history's first coat of armor, the felted wall of a ger, or a rug that can be walked on for hundreds of years.

I take the lamb from Gwen and pass her gently out the door, back with the others. She looks so much smaller than before but rounded and muscly, pearly white. The grazing season has treated her well. The other sheep sniff her as if they are meeting her for the first time.

Gwen pauses and watches the lamb's mother check her over. "She's saying, 'I'll still love you, even though you look ugly,'" Gwen says. We laugh. But to me she doesn't look ugly at all.

Zoe scoops up the fleece, rolls it into a bundle, and bags it. She's bagging and sweeping and helping me grab the next sheep from the pen. I'm sitting each sheep down and trimming its feet next to Gwen while she shears another one. We keep it rolling, one after another. Gwen makes the job look easy and elegant, but it's not easy to keep a mostly feral sheep propped calmly on its butt on a plywood board while you give it a pedicure. Every muscle in my body is working overtime. If we break stride it's usually because I lose my hold when a particularly bad-tempered sheep kicks violently and twists sideways to jump up. Years ago that very thing happened with a 150-pound ram whose hooves I was trimming. I kept going, but by the end of the day I had trouble walking and my left leg looked like it had elephantiasis; it turned out my quad ligament had torn away from my knee, which had me on crutches for weeks.

While we work, we catch up on news of other sheep farmers we both know. Gwen tells me about all the strange barters she gets (bottles of whiskey, whole pigs), the places she goes where she has a team of young and eager "agrarian hipsters" helping her like a pit crew, and other places where there's no one there, the sheep still out in the field for her to catch. She tells me about long days of work and of driving, and about falling asleep at the wheel last month in her beloved truck that had 430,000 miles, crashing through a ditch of knotweed to come to a stop inches in front of a telephone pole.

And we talk about our mothers. Gwen's mother still lives alone but has had dementia for ten years.

"She's the same, but worse," Gwen says when I ask. "A friend of my mother's who hadn't seen her for a long time stopped by the other day. Then she called me very distressed. 'Are you aware that your mother is wearing a dog leash for a belt?' she said. Oh, and also, 'We played Scrabble, and every single time she took new letters she had to ask me how many to take.' What am I supposed to tell her?"

John, who has just appeared in the doorway, adds in a lovely Irish lilt: "Yeah, I wanted to say, like what're ya doing eh, wanderin' the countryside with your game til ya find some poor demented lady ya can finally beat in Scrabble?"

My own mother has just begun to decline. She has always possessed a keen sense of social graces and proper language, so it isn't easy to tell when she is a little slower to grasp things. She simply knows how to fake the right thing. At first her disorientation was spatial—misplacing her keys or her car, or not understanding how to translate instructions from a page into the real thing. She gave up her photography and dark room, which was how we imagined her spending her elder years. As a photographer she had always been interested in reflections, patterns, the clarity of very early morning light. She took a portrait of me once, walking down a hill in a ghost town in Arizona, only my image is a reflection in the window of a junk shop, refracted by other reflections of road signs and trees and by trinkets dimly seen inside the glass. Now her mind, which has always loved complexity and mystery, is fading behind a dusty glass where we can't follow.

By the time the last sheep is shorn, the sun has come out and we're sticky with sweat and suint, a wonderful word for the greasy grime you get all over your clothes when you roll around with sheep for the better part of a day.

Zoe and I herd the sheep out of the paddock and past the gardens to the big pasture. They leap and kick their feet sideways as they run downhill through the gate, as if overjoyed to feel the cool wind on their naked backs. The sky has cleared, the sun glinting off the river and the wet road far below.

The bags of wool, lustrous and long in shades of sepia and silver, white and gold and deepest brown, are piled up in a bay near where the other harvest of summer—the grasses that also became this wool—is baled and stored.

We head inside for lunch, and I put out bowls of squash cooked with onion, carrots, curry, and coconut milk. There's also bread, manchego sheep's cheese, pesto, and a bowl of apples from our trees. The four of us sit at the kitchen table in the clear fall sunlight, the door propped open and the dogs coming and going underfoot like small children announcing imaginary games. They, like us, seem to sense a change in the sharpening air. Beyond the window a fleet of leaves sails south across the pond to crowd at the far edge, to sink and solidify in the layers, to move from air to water to the winter lair of fish and frogs. Skeins of geese drape low over the fields aiming for the river, wings tilting as they slow their flight, their talk incessant and laden with longing for where they are going, where they have been.

As I cut up a wormy apple, I'm aware of how my hands ache, and a bone-deep bruise made by a horn on the back of my thigh throbs against the chair. I'm aware of my own fear of getting hurt, and most of all of getting old. I'm afraid I'll be that old woman with moth-eaten clothes and a dog leash for a belt, eating moldy bread and forgetting my daughter's phone number, losing my shoes. I'm aware, too, of how happy it makes me to be in the company of these dear people, sharing this work as we have in a seasonal ritual for so many years.

Something about seeing your parents age can make you feel old, even when you're not. Even when your body still feels strong and you are just cranking through life, their infirmity can make you start to wonder how much time you have left and whether you are doing the right things with it. Even if I'm in pain, I want to be able to create things by hand until I die. I want to be a producer, not a consumer. I want to be like my English grandmother, knitting sweaters for everyone in the family and growing her own food into her nineties. Perhaps this is a fantasy. I did make a blanket with Wren, to continue the tradition, and I was proud that it was made from wool from our flock and is even more beautiful than mine. I wanted her to have it on her bed when she left home. But almost none of the dyes were made from our land, I didn't find the time to spin the wool myself, and Wren knitted almost every square. (Wren teases me that I unpick everything four times before I'm happy with it.) So, basically Wren made her blanket! But I don't give up hope.

Every year I go to the Fall Sheep and Wool Festival, and my mind is blown by the beautiful fleeces from the wildly diverse breeds of sheep, by the plant-dyed skeins lined up in a palette of earthy colors from the deepest indigo to palest apricot, by people spinning lace, making hand-carved spindles and buttons from sheep's horn. It's a world that reveres what is made the slow way, by hand, each piece one of a kind. I come home from the fair every year with a fleeting fever to raise more sheep with really unusual fiber, like Lincoln Longwool with its corkscrew curls, and to learn to spin again. But I'm a restless person and have a hard time sitting still; hence my perpetual dreams of walking with sheep across a mountain range. I'm waiting until I'm old and stiff enough to sit down and join that magical fiber club. Maybe it won't be long.

Gwen is eating with her good hand. Her other hand looks like she just pulled it out of a hornet's nest. I want to give her an ice pack, or a couch.

As we finish our lunch she asks me if I know the next farm she's going to, in Morrisville, but it's one I've never heard of. For her sake I hope it's easy to find on the maze of back roads in that town. She's going all the way there to shear two sheep. Tomorrow she will go to New York state to shear for four ten-hour days at a sheep dairy, a couple of hundred sheep a day, each sheep as tall as a Great Dane. It sounds brutal. If her hand is really injured, I wonder how she'll make it through.

"Do you ever think about a different career?" I ask as we clear the table.

"Sometimes I fantasize about being a librarian," Gwen says after a while. "Or maybe owning a flower shop. Something really quiet and gentle. But then again, I'd probably lose my mind."

"Yes," I say, "we would."

CHAPTER 9

GIFTS

> From time to time, he leaned over, he took some sky in his hands. It ran between his fingers.
>
> —JEAN GIONO, *THE SERPENT OF STARS*

I WRESTLE MYSELF OUT OF BED JUST BEFORE 5 A.M., PULL ON my jeans, shirt, and old blue sweatshirt from the heap of clothes on the floor where I can find them by feel in the dark. My socks and underwear are conveniently still in my jeans. I've been awake a long time, lying on my back and thinking about the day but not wanting it to begin. Not this one.

The dogs trundle down the stairs with me, and I quiet their excitement at the door. Machias, our shepherd puppy, can't pass any threshold without a loud woof. I make tea in a thermal mug, pull on my barn boots and canvas coat, and head out. The dogs run off to the east to patrol the field, as they do every morning, Machias barking with excitement tinged with fear. He wants to make sure that whatever might be out there in the darkness knows he's coming. Rue's huntress instinct hasn't dimmed, but her old legs are stiff and her gait no longer effortless. She runs behind, which I know hurts her dignity. She's the boss.

A pale-yellow light rims the hills below a scattering of stars. Frost glints in the grass. I hear a barred owl calling from the woods above the fields as I walk out around the house toward the road. In a few minutes the dogs return, and I coax them into the back of the truck. Rue has decided she's a cab dog and looks at me with longing eyes.

"C'mon old girl." I pat the tailgate. "You can do it. We're not going far."

We head up Bragg Hill, truck lights bouncing over the ruts and lighting up the trees on either side of the road—orange maples with their inner light and the somber burnish of ash like spadefuls of ochre clay. We pass open fields full of huge round hay bales sleeping in the dark. Five deer raise their heads, then dash toward the woods, white tails flashing in my headlights. Soon it will be hunting season. The deer look sleek and fat.

My thoughts feel slow, my belly hollow in a way that food won't fill, hollow as if I just missed a step on the stairs in the dark.

At the top of Bragg Hill I park my truck, and the dogs and I hop into Bowen and Charlie's truck, already hitched to their horse trailer and parked for me alongside the road by their barn. I feel small in the wide leather seat and have to slide forward to reach the pedals. But it's an automatic and practically drives itself compared to the truck I'm used to, which needs to be jammed into gear with ambition. I glide down the hill in silence, remembering to make a wide turn into our narrow farm road.

The first time I borrowed the trailer I cut the corner too sharp and hung the trailer up on a high ditch. I couldn't move in either direction. It was night, beginning to rain. Ky, our neighbor and a person I'd trust to save my life, was driving home and dropped whatever he was doing to help out. We spent a dark evening blocking and rocking and winching the trailer off the bank. Ky had all the right tools, was never in a hurry, and never said anything that hinted at my stupidity. Since then, I'm a better trailer driver, but still a nervous one. I want to honor the kindness of neighbors by making sure I don't screw up.

Out in the barn, the young rams wait for me in a pen in the hay-scented darkness, resting their chins on each other's backs, blowing white puffs of breath through flared nostrils. I flick on some lights—the bulbs dim in their dusty cobwebbed cages by the ceiling—and grab a bucket to give the animals a last drink before they have to go.

Until last night when I rounded them up, these sheep have spent nearly every day of their eight-month lives trailing after their mothers though long meadow grasses heavy with dew, bleating when their mothers got too far ahead, gradually gaining enough confidence to play together at the edges of sight, testing their newfound daring. Any unfamiliar sound sends them pronking back to the flock that is their food, their safety, their everything.

Most of these young sheep came into the world as a double and are never far from their twin, whom they know from all the other lambs by sight and smell and possibly a million other things we don't understand. Even now at eight months, sexually mature, nearly grown—the equivalent of young adults—they return each night to their sibling to sleep touching one another. "We were as twinn'd lambs that did frisk i' the sun," my father likes to quote Polixenes, the part he played in *The Winter's Tale* as a young man in university in London. Shakespeare knew, as did his audience, that twin lambs have a special bond. To allow animals their natural familial bonding and to let them wean themselves from their mother's milk only when they are ready (or she is!) is, to me, one of the most beautiful parts of the pastoral system we practice.

The ram lambs that I don't keep for breeding my own flock go to other farms as studs. Most, but not all. Some haven't grown as well or have close-set horns. Some have an unknown sire or simply don't match what people are looking for that year. If I kept all the lambs that I wanted each year, my flock would soon exceed the ability of my land to support it, so these nearly grown rams are going to the slaughterhouse today.

Overstocking causes the well-being of everything—land and animals—to decline. But careful discipline and awareness of the whole system has often been ignored, and it was circumvented entirely when we invented the modern system of raising ruminants for meat in confinement feed lots, with no grazing at all, bringing in corn and soy from vast monocropped fields elsewhere. Confined animals are not "overgrazing." But they are not healthy either, which is why we lace their food with antibiotics and hormones. Similarly, the land gets its dose of chemicals to encourage it to do what it was always meant to do naturally but is now too depleted to do alone.

We can now eat meat with no connection to the whole picture of where meat comes from. A week ago, anticipating my freezer at the farm store being full of meat, I posted an announcement for "Local Grass-Fed Lamb for Sale" on our local online forum (Front Porch Forum, or FPF). The next day I saw this public reply:

"I would be thrilled if posting dead baby animals for sale were not allowed, so we could read FPF without feeling disgust or sadness."

It seemed that this woman's distress came not only from the word *lamb* being used to refer to meat but the whole idea that her neighbor might be involved in butchering. Didn't meat come from the supermarket?

Another neighbor responded to her post: "I completely agree! Some people are really offended by this. The valley is made up of many people and some of us have moved here from other states where this is not part of our culture."

Exactly. It's hard to move to a rural place and not be judgmental of the farms, whose muck and stink and death are often right out there for everyone to see. It's uncomfortable, so the farmers should change. Be a petting zoo. Sell pumpkins. Or at least stop showing us where our food comes from.

I know I sound cynical, and not everyone deserved the rant I wrote back on Front Porch Forum. Farming long enough—tiring yourself out while others think it's either unsavory or romantic, when the truth is that it's hard and complex and individual, and how should they know unless they experience it themselves, which they won't—can make you a bit cynical. I've been called Little Bo-Peep by visitors. I've been called a murderer when I try to sell our meat. I don't feel like either of those things.

But if I step back, feel more generous and humble, I notice that the way people who are not farmers think about farming is not unlike the way most of us approach nature: It's either a beautiful backdrop, a screensaver, a playlist of soothing sleep sounds, or it's full of frightening hazards and germs and grime, to be struck down or controlled. Nature—and by extension the pastoral farm—exists largely in our imaginations as a place of wholeness and health, purity, and even virtue, or as a place that's dirty and uncomfortable, something to be avoided.

Either approach is self-referential. Western culture gives us stories of "pure" nature, without people—or sometimes with Indigenous people, who are roped into the same fantasy—that are meaningful not in themselves but as a redemption to human destructiveness, or as a place of spiritual revelation. On the other end of the spectrum, a revulsion of nature can be much more than a response to how modern life has produced an unfamiliarity with dirt, bugs, and hard manual labor. It can be a response to ancestral trauma on the land and a need to find distance from the oppression of forced labor. Whether we learned to avoid the natural world because of trauma and legitimate fear or became conditioned to avoid it through affluence and notions of material progress, the result is the same. Most of us, including me, end up with some narrow definitions of what nature means, and they all in some way relate back to the human. It has become my project to try to see nature—and by extension, farming—as it is, and to have a deep relationship with that realness. No doubt I fail, but the attempt feels meaningful.

PETER DUCKS THROUGH THE LOW BARN DOOR WITH A STEAMing mug of coffee. The sky is almost light now, and he's silhouetted against clear pale blue.

He sees the long look on my face and flashes a smile.

"You ready for this?"

"Never. But let's do it."

I've reluctantly chosen several ewe lambs to join the truck with the rams, twenty in all. These lambs are big now, not quite fully grown but heavier than I can lift: nearly a hundred pounds of muscle and bone, long silky fleece, and curled horn. One by one we catch the lambs and walk them out the door and up a grassy slope to the farm road where I parked the trailer. The sheep brace their pointy feet into the sod as we walk them, one of us on each side holding a horn and pulling them forward. We lift them into the horse trailer, sliding the gate across each time with a clang. The sheep are nervous, but once inside they begin to eat the hay I have piled at the far end. As we load them, my biggest consolation is knowing they've had a wonderful life.

I say goodbye to Peter and the dogs and switchback my way over the Appalachian Gap, the highest road in the state, as the sun comes up over my shoulder and the sky deepens to a pure autumnal blue. I figure out how to work the fancy Bluetooth audio and settle in for the long drive.

At the top of the Gap I cross the snow line and pass through a fringe of frosted fir trees and bare birches, briefly traversing a world of winter before descending back into the red and gold of autumn on the other side.

I stop at the bagel shop in Bristol for a coffee. I climb onto the wheel hub to check on the animals through vents in the upper side wall. Most are chewing their cuds, the wind ruffling their long fleeces from the open windows. A gray ram curls his lips back, getting ready to mount a ewe who instinctively squats to pee.

I continue west to the very edge of the state, turning south along the embankment of Lake Champlain where the land is flat and fertile, the sky weaving its glittering thread over the bright water all the way to New York's Adirondacks. Huge dairy farms and cornfields stretch on either side of the road; this is Addison County, the heart of Vermont's farm country, though the cow dairies have been declining and going bankrupt for years, and many bear the look of resignation. Today the farms are quiet, the cows inside their massive free stall barns, the corn harvested and stored as silage for winter feed, mounds of plastic-wrapped round bales stacked by the barns and along the fields. I pass a farmer spraying liquid manure on a field of corn stubble, and for a while the road is slick with fresh cow shit and the air musky and sharp.

When I'm nearly at the southern tip of the lake, I turn right at a taco stand that looks as if it never opened for business, pass an elementary school, and enter Benson, a blip of a town at a crossroads with a tiny inn and a general store. I turn left and rattle along a dirt track past fields full of beef cattle to Hathaway Farm, which is also a federally inspected slaughterhouse and butcher shop. They raise their own beef and pork and take customers from all over the Northeast. Most of their workers are Guatemalan, here on ag-worker visas.

I park the trailer by the porch and get out, stretch my back, yawn, do a little shudder. I feel like I do just before I'm about to run a race. I go inside to find Annie, the woman in the office who runs this show. Beside the stairs up to her office are heavy steel doors leading to the cutting rooms, and beyond that, the kill floor where the animals are skinned and gutted, then hung for a few days before being cut and packaged. I can just make out the meat cutters through the panes of foggy glass, dressed in long gowns, blue gloves, and head coverings, like surgeons. The cold air smells strongly of bleach and blood.

Upstairs Annie sits under a large half-moon-shaped window, surrounded by stacks of flattened cardboard boxes, bins of aromatic spices for mixing sausage, and rolls of meat labels with the names and logos of dozens of farms—Snug Valley, Happy Acres, Boyden Family Farm. Most raise grass-fed beef. She's on the phone, with an unhappy customer it sounds like. This is not uncommon, I know from talking to Annie. There is a lot that can be lost in translation between a farmer's expectations of the amount of meat or cuts they will get back from any given animal and what is actually possible from the butcher's end of things. Annie spins her chair around to signal to me to come back later.

Craig, an older man with a knife holster cinched around his angular hips, a camo baseball cap, and a cigarette dangling between his lips, meets me in the yard and guides me as I back up the trailer to a narrow concrete hallway between two barns. After pulling forward and straightening the trailer for the fourth time, I get it good enough.

A tall white ram lamb gingerly steps off the metal gate into the barn alley. The rest follow, jostling and flowing like milky water into a giant shed with twenty-foot gates that clang shut behind them. Craig opens and closes a few partitions to section my sheep off from a group of mean-eyed spotted pigs on one side and a few leggy Katahdins on the other. The latter are a breed popular for meat because they are "hair sheep" and don't require shearing. If you're in the meat business, dealing with wool is just a nuisance.

My sheep huddle tightly together at one end of their stall, the rams no longer jostling to mount the ewes as they did in the trailer. They stand in a thin layer of muck over the concrete. Just on the other side of the wall are large plastic dumpsters filled with offal—guts, heads, feet—stained with blood, swarming with blue bottle flies. Someone comes daily to pick them up and empty them somewhere, but I don't know where. Even I don't have the whole picture of this food-production process.

"You'll give them hay and water, right?" I say to Craig.

"Water, yup. They go on the floor as soon as we're done with these pigs."

I stand in the alley after Craig heads back into the barn through the chute where each of my animals will be led in and have a fatal encounter with a stun gun before entering the slaughterhouse kill floor. The stun gun, which has no explosion or bullet, goes into the brain and is the most humane, immediate way to kill an animal. I've watched Mary do it on farms where she gives farmers the chance to harvest their own meat right at home, and she's tender and soothing to the animals, right up to the end. But I don't usually stick around to watch it here. I bring the animals here because I want the meat to be cut and inspected to sell to stores, but I also come because slaughtering this many animals myself would be too much for me. "Killing animals is really heavy and scary," Mary said to me once when I rode along with her to a farm to see her in action. "When I started, the adrenaline was too much. I had to take baby steps to memorize every move. It's like a whole dance."

I look through the gates at my animals. Cool air billows into the shed from the open sky, and all of the animals stand quietly. One of my rams starts to chew his cud.

I can't help wishing I had kept one ewe lamb in particular—a spotted badgerface with beautiful markings. Too late now. Alone for a minute, I reach through the rusty orange bars of the pen to hold a white ram lamb's woolly cheeks between my hands. He has a burr buried in the velvety fleece by his ear, and I tease it out. He puffs fermented grass breath at me, blinks, lets out a belch. "Thank you," I whisper. The hollow in my stomach expands into my chest. I've done this so many times, and it never gets easy.

Greg Hathaway, the slaughterhouse owner, comes out of the sterile-smelling processing room in a billow of ice fog and a waft of disinfectant, his bright, round face chapped by being perpetually cold.

"Hey. How's it going?" I ask. I force my voice to sound strong. Everything's cool.

"Terrible," he says. "We lost power in that last storm, and the generator broke—lost some of our customer's meat before we could get it back and running."

"I'm sorry, that sucks. I saw Annie trying to calm someone down on the phone."

"Yup. Lots of that going on. You bringing more damn sheep today?"

It was half complaint, half a tease: Any chance he gets he complains about how he hates to butcher sheep, how much work it is with all those little bones, no money in it. But when he decided he wasn't taking any more sheep customers, he kindly allowed me and a few other people to stay on. I was relieved. It's almost impossible to get a date anywhere given the limited number of slaughterhouses in the state, and even with the Hathaways, I have to book my dates a year in advance. Butchers charge by the animal—$145 per sheep—which puts a burden on the farmer to raise an animal to a good weight, so you can still make some money on the amount of meat you get back. This is very difficult with grass-fed sheep. My sheep are also a "primitive" breed, which means that they can only breed once, in the autumn, to have lambs in the spring. Some breeds can be bred in fall and spring, which allows the farmer to have lamb for the spring Easter market or to spread out lambing to have meat year-round. In contrast, my meat is a seasonal specialty.

At my scale, particulars really matter: what my breed of sheep are good for, how I raise them, the quality of their meat and wool, and my attention to track and improve their breeding lines all make it possible for me to make a slim livelihood. I am not, I realize, a grass farmer quite like Greg's other customers who cultivate good grassland only to produce meat. I am a pastoralist, raising animals for a huge range of reasons, meat only a small part of it. Our diet is also pastoral; we eat our own meat but not daily, or even weekly. If we all ate meat at special occasions or as a flavorful and healthy garnish rather than a staple, the industrial meat industry would have to shrink. I think that would be a good thing.

Greg and I talk awhile in the gravel parking lot near the animal sheds. I tell him about the early snow on the mountain pass.

"You can keep it. Thirty-eight years, and I still hate winter," he says.

The late-morning rays warm the frost on the north slope of the roof, and water splashes down onto pitted concrete in the sunlight. A huge weeping willow sheds its golden leaves in the yard. Across the fields, I can see Greg's pigs rooting up the ground with their snouts, stirring mounds of soil, compost, and rotting hay, speeding up the life-giving force of earthly decay.

A young couple drives in with a single cow tied with a halter in the back of a Ford pickup with handmade wooden sides. The cow looks like an old family milker, with humped withers and one crooked horn. The whole getup makes me grateful for living somewhere with a culture that resists efficiency, that has a human scale, where people make the most of what they have. This, too, is pastoralism. Pastoralists trade efficiency for relationship, for an innate sense of reciprocity with land, animals, and people. Small farms in Vermont survive and even thrive because more and more people (my FPF friend not included) value relationship over efficiency and will go out of their way to come to places like our farm for our meat, because they care about us and our story. As long as this ethic lasts, we'll be able to make a living; without it, we won't.

I HEAD BACK INSIDE TO THE OFFICE TO SEE ANNIE. I NEED TO fill out a sheet of instructions for the meat cutters to tell how I want everything packed for my customers: what thickness of chop, bone-in or boneless leg, shanks or grind, what sausage spices. Next week I will drive back here and pick up boxes packed full of meat, all vacuum-packed, frozen, and labeled by cut and weight.

The whole thing feels like Narnia to me. I drop off my lambs, their souls or animal selves enter another world, and I come back to the wardrobe and see just a set of clothes, nothing more. It's weird, and I feel its weirdness, and yet I have found ways to reconcile it. I see my job—not to prevent death, but to try to prevent suffering. I have seen so many natural deaths on the farm that were so much more full of suffering than what will happen here in a few hours by stun gun as they are chewing their cud. I do appreciate this beautiful food that I know is from grass and rain and sun and my love. I do get excited to go home and share this food with appreciative customers.

ANNIE SWIVELS HER DESK CHAIR AROUND AS I COME UP THE stairs. She and I have gotten close over the years. Annie is one of those people whose age is impossible to guess. Time has worn her down, slumped her cheeks and hooded her eyes, eroded wrinkles into every inch of her face. She sighs whenever she sees me, and I listen to her worries.

Today was rough, she says, because she has all these unhappy customers, and on top of that, her pet goat has just died.

"I'm not sure what to do. My other goat is all alone now and just cries all the time."

I am sitting at a folding table underneath posters of animals drawn and quartered, showing all the possible ways to butcher. I'm trying to decide if I want "neck slices" this time or "grind."

"Maybe you need a sheep," I hear myself say.

"What do you mean? One of yours?"

I look up, put down my pen. "Yes. Just take one. I mean, I didn't want to bring them all anyway. There are some sweet females."

In a matter of three minutes it's decided. I barter a couple of years of meat labels, which Annie prints for me at home as a side hustle, for a sheep.

I finish filling out my cut sheets and follow Annie outside. I watch as she goes to the shed with Craig, points without hesitation to the badgerface ewe lamb with brown-and-black patterns circling her eyes.

"That one. I'm going to take her home," she says to Craig, who looks like this is the craziest thing that's happened in a long time. Without a word, he walks the ewe out of the vast concrete waiting pen, past the kill chute, back into the rich autumn light of the yard. We lift her into the back of Annie's Subaru while Annie waits eagerly at the wheel. She and her lonely goat live just down the road.

CHAPTER 10

TRACES

It seemed to me that the days had run from everywhere like water through a basket.

—JEAN GIONO, *THE SERPENT OF STARS*

Before I leave the farm for the day to pick up my mother and bring her to our house for Thanksgiving, I need to escort my ewe lambs to their own field, safely two pastures away from the rams in rut. A young ewe, called a gimmer, who is bred her first fall at six months of age will always stay small and is the most likely to have trouble in delivery or to reject her lamb. Although they are sexually mature, I like to prevent them from breeding and to watch them grow tall and fat during their second summer without young ones to feed. But it's not easy; the rams will break down gates to mate with them.

   The day starts gray and cold, the sky like clabbered milk. I open the east gate, and the fifteen ewe lambs I have chosen to keep this year scatter out across the frozen field, not sure where to go without the older ewes to lead them. I run to get in front of them and then push them to the right, downhill, toward the south gate. They will spend the next two months in a field I have set up around a hawthorn tree. The tree itself is broad and low, with the shape and thorny character of an acacia. It's a favorite place for the flock, and over many years they have congregated there, for shade or shelter from the rain, wearing a disk of bare earth around the gnarled roots and out to the drip line of the branches. They have nibbled as high as they can reach on the tree's lower limbs so that its profile is perfectly even around the edge, a domed parasol against the sky. Like threads fanned out from the circle of understory, little trailways made by the sheep's hooves squiggle through the grass and disappear into the larger tapestry of the field, evidence of their tendency to follow, one in front of the other, the path of the familiar.

 I shut the gate, and the sheep begin to wander the perimeter of their new domain, nibbling on the frozen clumps of grass poking out of the snow. They are old enough to be separated from their mothers, but it feels unfamiliar to them. For a few days they will gather at the gate, not quite understanding where they are, knowing something is missing.

MY MOTHER RECENTLY MOVED FROM HER HOME ON A mountainous, rutted dirt road to a perfectly tidy assisted-living community in a small town, a parking lot away from the hospital that pinned her femur and mended her broken elbow after she fell on the road outside her house. After surgery and two weeks of rehab in the hospital, she never went home again. Where she is now she's not quite sure, that much is clear. Being in the hospital mended her broken body but seriously tilted her mind. We are learning to navigate by the stars.

 I pull into the parking lot at the retirement home, and there she is, already outside. Ready. Possibly packed since yesterday, when I called to remind her I was coming. Hair to rival Einstein, cheeks puffed from medication and blooming red with a fine net of veins, eyes so bright they would appear maniacal if not so trusting and kind, a smile that crinkles her entire face, and a walk like a penguin balancing an egg on its feet, as if her legs are made of wood.

"Look at meeeeee," she declares as she careens off-kilter around the car to where I'm holding the door open for her. "I'm walking. *Can* you believe it!"

And I can't, really. The last time I saw her she was hobbling around with a walker, and the scar down her leg looked like the bark of an ancient oak had taken the place of her skin. This time she doesn't even have her cane.

As I drive home, I tell her about my recent trip to England with Wren and my father, to visit relatives, including Mom's sister in London. My mother was the only one in her family ever to leave England, emigrating to America with my father when they were in their twenties. Her trips "home," once frequent, are rare now.

"It was so green," I say as we drive north through the steel-gray and snow-dusted hills of a Vermont November. "The parks in London were so beautiful."

"Were there lots of daffodils?" she asks, and I hesitate, trying to understand where her mind went just then, as it's fall, even in London. Perhaps, in her mental equation, "so green" = spring = daffodils. My mother's mind, more and more, hops aboard trains that run on a different kind of logic.

I quickly learned not to correct these odd leaps—which would just add to her confusion—but instead to board the same train she was on and ride along. It's a game of peering into an imaginary landscape with traces of what's familiar from my own long life with her and traces of something only she sees, conjured from deeper memories or dreams. We can still have wonderful conversations this way, as we have all my life, and the details hardly seem to matter.

"I didn't see any daffodils," I say. "Do you remember, though, all the millions of flowers lining the Mall after Princess Diana died?"

"Yes," she says. "Yes, I most certainly do."

THE FIRST SIGN OF MY MOTHER'S MIND GOING WAS BEFORE HER recent fall and her rehabilitation in the hospital, though it's clear that this event disoriented her badly. Some time before, without warning, she reverted to driving on the left side of the road. The first time it happened she was with my nephew, and he said it wasn't as if she'd drifted across the road, more like she firmly planted herself in the wrong lane and seemed surprised when he got alarmed. The next time, she was driving alone and caused a head-on collision on the twisty back roads near her house. Miraculously no one died, but that was it for her driving.

The only explanation I had for this new habit of my mother's was that it came from a very old habit. She had learned to drive on the left in England, before moving to the States. Although she had been driving on the right side of the road four times longer than she had driven on the left, it was as if those early memories grew brighter as her mind began to dim. I say memories, but it isn't that, exactly. She wasn't remembering her days of driving on the left, she was just doing it, on autopilot, as a free association. The equivalent for me was walking into the house where I spent my entire childhood and reaching for the light switch on the wall where it was when I was small, even though it hadn't been there for years. More and more, those kinds of associations—the way one thing is linked to another in the consciousness, but outside the realm of logic—seem to be what stay with my mother the strongest as so much falls away.

By the time we pull into Knoll Farm and I settle Mom into a chair by the woodstove, everyone has gathered in the kitchen of our old rambling farmhouse, once an inn and full of small rooms and an overabundance of doors. Our kitchen alone had six doors leading out of it when we moved here—two to the outdoors, two to different mud rooms, and two to other rooms in the house. When the girls were small, playing on the floor, we could always feel the wind blowing under the doors and across the room. It felt like a cold alley, which prompted us to sheetrock over two of the doors and knock down a wall. The woodstove—a huge barrel with a graceful iron Dutch elm on the front, arching its limbs over a Pyrex pie plate for a door—did its job to counter any lingering drafts.

It's hours before Thanksgiving dinner, but everyone is crammed into the one important room, which always makes me wonder why anyone needs a house bigger than a kitchen with a loft for a bed. Peter's brother, Glenn, wanders around, his small blind poodle, Lucy, under one arm to shield her from Rue, who would love to herd her into the barn and make sure she stays there. Glenn's wife, Jamie, sits down on the arm of my mother's chair, near the woodstove, and listens to her explain why she stopped driving.

"It wasn't going so well," Mom admits, always the master of understatement. "Everyone was quite relieved when I offered to give it up," she says with a chuckle. No dementia there.

The crush of people in the kitchen delights my mother but also confuses her. I can tell from her face that she's trying hard to track the conversation and figure out how all these people are related to her, or at least to me. "Now, you are . . . ?" is the way she begins each new interaction. She wants to set the table, and Peter starts to help her, but she goes automatically to each drawer and shelf, setting each place at the long table the English way, with the spoon above the plate at twelve o'clock.

We have leg of lamb for Thanksgiving, boned and rubbed with ginger, garlic, star anise, hoisin sauce, and cumin, then rolled and tied with string. I slice it thin, and the red juices run out and mingle on the serving plate with the salty crust, roasted cloves of garlic, and charred bits of onion gleaming with oil. We have roasted potatoes, leeks braised with thyme and cream, brussels sprouts, and butternut squash, all from the farm. I can't describe the profound gratitude and pride I feel to share this food.

The morning after Thanksgiving, Peter's family is packing up to leave, talking about how long the drive back to New York will take them given the traffic near the city. For a couple of days Mom will stay with us. I have explained this to her half a dozen times or more and even heard her repeat it to Peter the night before. And yet, a few minutes after everyone else leaves, Mom reappears in the kitchen with her bags packed and her blue parka and sneakers on.

"Right then," she declares. "We're off!"

If the memory is like the many layers of an onion, as I imagine it, time is the outer layer, and the first translucent skin to shrink and unravel. The concept of *tomorrow* or the exact location in the future of *Friday* no longer seem to have meaning to my mother's mind, no matter how many times you explain it.

Back in the bedroom, we sit on the bed and I open her case, which she keeps trying to zip closed so that she's ready when I am. "It's okay," I say. "I know everyone is leaving, but you are here tonight, then one more night, then I'll take you back. You'll have more fun with us."

Always agreeable, she says, "Oh! Well that's fine, then," as if I'm telling her for the first time. She keeps trying to lean back on the bed as if she were on the couch, which is usually where we talk, and so she cantilevers back, ever so slowly, then rights herself when there's nothing to lean up against. A few minutes later she tries again, leaning ever so slowly back and wondering why nothing is there.

"Do you know what might help you with the time thing? Maybe if you hang up a calendar and can mark which day. . ." I'm reaching. I have no idea what helps with this, if anything.

"Does it feel hard when you get it mixed up?" I ask.

"You know, this really is the one thing that makes me feel genuinely crazy," she says, and she looks upset, a look I see rarely on my mother's face. "In one ear and out the other. You could tell me when we're going a million times, and it just won't stick."

After everyone leaves, Peter and I make lunch and Mom sits at the table shelling some scarlet runner beans I collected from the garden in September and had drying by the stove. She pops open the long brown husks to scoop out the shiny pink-speckled black beans inside, and her hands work like they always did at this familiar task, without the need for thought. I think of her, her mother visiting from England, and me and Kate, all squatting on the cool slate of the farmhouse porch where I grew up, shelling peas until our thumbs turned green.

I HEAD OUT TO THE BARN TO FEED THE SHEEP. THE CHICKENS are keeping their feet warm by sitting on the sheep's woolly backs, and with my arrival the sheep shrug them off, as if embarrassed for me to see. It feels good to walk, freely, the cold air slapping my face.

After feeding the rams and breeding ewes, I walk out to the ewe lambs, who are huddled at the gate but follow me toward the tree as I drag the big sled of hay by leaning into the thick rope wrapped across my chest. As I walk in their trails to find smoother footing, my heavy sled bumping over the lumpy frozen ground, an easy metaphor emerges. The animal paths suddenly appear to me as fractals of the nerve endings in our minds, some more firmly trodden and worn than others and therefore smoother and easier to repeat, and some trailing off, not quite connected to anything else. Though clearly dementia, and just plain old age, changes the rules of the game, it's true that the more we practice a way of thinking, or a physical skill, the stronger that nerve pathway gets and the harder it is to lose.

My mother often says, "If only I could be useful, that would make me happy." After a life spent creating a homestead in rural New Hampshire where our family possessed an almost absurd zeal for self-sufficiency—heating with wood, growing most of our own food, milking cows, making butter and cheese, boiling sap, dying wool—my mother hates sitting and not being able to contribute. Here on the farm, there's almost always something I can give her to do with her hands that her body remembers instantly how to do.

There is a place for elders to find meaningful work in an agrarian culture. In our industrialized culture, so concerned with efficiency and safety, we shut our elders away in a sanitary place without any connection to the world of work, of weather, of things being made and born. Not only that, but most young people grow up occupying themselves primarily with their minds—with technology—rather than with their hands, learning to make and to do things. What will be the equivalent for today's young, when they are really old, of something as hardwired to my mother as sitting on the porch shelling peas? What will be, to use William Bryant Logan's phrase, the ways we possess the "knowledge of objective truth"? Imagine watching someone really experienced use a tool—a blacksmith say, or a woodworker with a hand plane. The way they move the tool reveals the thinking not just in their heads but in their hands, a kind of truth that comes from years of minute perception and intuition. Logan writes: "It is not thinking. It is not feeling. It is the training of perception in the face of resistant materials. It is need and thought adjusting to the real. It is a new participation in the world of relationship, one through which a strange kind of flexible precision emerges unbidden."

When my mother pauses, as she often does now, to search for a word, it's as if she's there, at the end of one faint sheep path where it fades back into the field, not sure how to cross the blank space to pick up the next.

I wonder what is more precious: These paths in the physical world that forever etch inside us to guide our bodies, like our animal kin, toward food or water or the gathering places for love, or the trails in our minds, those elusive words that are translated from objects we can see and touch to something potent yet ethereal, akin to poetry—our memory.

What will we each have to hold on to that our bodies remember how to do when our minds can no longer lead us there? What does it mean that our culture is so intent on preserving the ageless mind when at the same time it teaches us so little to do with our hands, our bodies, our hearts?

WHEN I COME INSIDE, WREN SAYS SHE TOOK MY MOM upstairs to help her have a bath and thinks I should go up and check on her. I find Mom sitting on the edge of the tub, her Depends partway up her soft thighs and her tights caught on her leg by the terrible scar, which looks sharp enough to rip fabric. The skin on the bottom of her legs looks so hard and scaly and purple from poor circulation that I want to cry.

"Look, Mom," I say. "Let me try some of this amazing skin cream on you that Peter gave me." I start rubbing her shins with cream as thick as butter and smelling strongly of lavender. The smell gives me an instant vision of my grandparents' tiny flat and Gran's linen closet with the child-sized door handle with a button in the center that you clicked with your thumb. Inside were little lacy lavender sachets she sewed from stems collected in her garden.

I keep rubbing, but the cream stays on the surface of Mom's skin like paint on a shingle.

She looks down at her legs, and it's almost as if she doesn't recognize them anymore, isn't bothered that they look like something from the zombie apocalypse.

"Now that was pure luxury," she says. "You've given me a real treat, Lovey. And here I am, looking like God only knows. God's last gasp, that's what."

The one thing about Mom that is utterly magic and never changes is how easy it is to make her absolutely happy and overflowing with appreciation; she's a bright-eyed sparrow with a blighted leg who always finds the next delicious crumb. Maybe, to her, a crumb is not even a crumb, but a whole cake. I have heard that the temperament one carries in the world, whether we are born with it or cultivate it, shines through when the mind is gone. True or not, the core of my mother is this brightness, a brimming glass.

When I take her back to her apartment the next day, we are both pretty quiet in the car. I miss our long talks that usually ended up with her giving me really good advice about something I was wrestling with, and I realize I don't initiate them as much anymore. I'm trying to get used to a different version of the intimacy we've had, trying not to share thoughts or news that would worry or confuse her. And yet it would be utterly foreign to be superficial with my mom. I settle for silence, and she seems content, drifting in and out of sleep. And I drive, already missing the woman sitting next to me.

"I hope you have a lovely holiday," she says, squeezing my arm and breaking the silence as we turn into her hospital-home complex. I know she is saying it because she means it, but also because it's what the brain comes up with when the body readies itself to leave. She is good at all these social graces. They haven't left her; she reads the cues even when she doesn't quite know what she's saying.

"Thank you. But I'll be with you!" I say, hugging her.

"Will you? Oh, well then!" she laughs, and I help her to the door.

"I love you," I say.

"I love you too, my darling. Come back soon."

We walk into the apartment complex, which she can never remember the name of but doesn't call "home." "That place I'm staying" is the closest she gets. I'm reminded of how very brave she is, to be here in a place that she knows will be her last, and not where she would have ever wanted to be, to be alone and lost in a way that isn't chosen, that isn't about coming home again. Her bravery undoes me, and I am overcome with sadness, with love, and with guilt.

We walk arm in arm to door 219 down the carpeted hall, which is lined with photos of the dairy and wheat farm that once thrived on the hillside where the hospital and parking lot and low line of retirement homes called Harvest Hill (honestly) were built. Her apartment is small, with a gray-brown rug and a beige couch, gauzy curtains over the windows. I help her settle in to rest, make her some tea. I give her a last hug and peel myself away.

As I close the door, I leave her a wish—that from her chair, her hands idle in her lap, her mind will travel into fields she has known, to the wide open, where the sheep go to rest when they are unafraid.

CHAPTER 11

DAMAGE AND HEALING

> Lost, rolled along in the herd like a bit of gravel, I held myself together around this shepherd's love.
>
> —JEAN GIONO, *THE SERPENT OF STARS*

THE OLD WHITE RAM, HIS FLEECE AS THICK AS A DUVET and hanging to the ground from his massive body, is shivering all over like a small naked boy who has played too long in the lake. But he isn't wet, and it isn't a cold day. Today is clear and strangely mild: the coming of a January thaw.

This ram, Ovid, is trembling from some internal battle, and I am to blame.

The day before, I moved Ovid out of a paddock with his group of breeding ewes, now all pregnant, and down into a winter field with three other rams. I looped a rope halter over his massive head and led him through the barn and down to the lower field along a snaking shoveled path through the snow. The banks kept him pinned. I held him with one hand, and with the other I tempted him along with a bucket of grain to the gate. I let him go into a field where there was an old pig shelter, a stream to drink from, a pile of hay. The other rams came over to greet him, and that was that. Or so I thought.

The four rams in the field were ones I had picked out for specific reasons to breed with a specific group of ewes. One had a recessive color gene I wanted to encourage in the flock, another was broad and stocky, another was from a maternal line of excellent milky ewes and the fourth, still small and immature, brought completely new genetics to the flock. Over the years I've bred dozens of rams, but I have my favorites, whose lines run through all my best sheep and who I remember for their distinct look and personality.

Every January after breeding season the rams meet again, coming back together after three months of cavorting with the females I chose for them. They are ragged from the effort of obeying their hormones, bony under their thick coats. During the rut the rams take little interest in food. Each stud chases his group of females around the enclosure, curling his lips back to sniff for estrus and putting his nose into the stream of urine when the females squat to pee. When a ewe in heat stands still for attention, doe-eyed and limp, the ram nickers along her flank with his lips and eventually mounts her from behind, gripping her belly with his front legs. It lasts but a minute, but he will mount her many times over the two days she is in heat. If she doesn't conceive, it will be another seventeen days before she cycles again. While a ewe is in heat, the ram won't leave her side. Sometimes he will guard her, keeping her in a corner of the paddock. When all his ewes are pregnant and no longer cycling, he will pace the fence lines, trolling for a ewe in estrus in a group on the other side. The farm gates between enclosures are pounded to flat ribbons from years of ram heads mashing against them.

For primitive breeds of sheep like ours, the flush of mating hormones is triggered by the days getting shorter in the fall. If I ran one ram with the ewes all year, he would breed them during their first cycle in November, and the ewes would give birth in April, and this would be easiest of all. But our hillside is small, and we can't run enough sheep to make a go of it with meat, so we improve the flock by selective breeding. Most of our animals go to other farms in diverse groups as "starter flocks" of a heritage breed. I keep my ram lambs "intact" (meaning they get to keep their testicles) partly out of kindness, but also so that I can see who grows well and sell the best for breeding. This makes managing the flock harder in the fall—with groups of rams and ewes dotted around in different enclosures, all of whom need to be hauled hay and water through the deep snow, and the rams doing constant damage to the gates and fences—but it's worth it. Given that each ram will be bred to many ewes, most of which will have twins, choosing a good ram does more than anything else to raise the quality of the flock.

As my eye for rams got more consistent over the years, people who were interested in my breed started to say that my flock had its own look, as do the flocks of other shepherds who get into selective breeding. Without really planning it, the conformation and coloring of the animals start to reflect the aesthetics of the owner as well as the specific animal adaptations to that land. After twenty years my sheep flock is gentler, calmer, taller, with an array of recessive fleece colors and patterns it didn't have before. They also have more parasite resistance and are adapted to thrive on a grass-only diet. There is an entangled pride in this that I can't explain. It seems like trying to manipulate something nature gives us, and yet it also links me all the way back to the first pastoralists ten thousand years ago or more who brought the wild Asiatic mouflons from the mountains into the narrow enclosed streets and courtyards of the village, breeding those they valued for their vigor and soft fleece, and probably also those they fell in love with for their sweetness and beauty.

The practice of carefully breeding the next generation of sheep for this farm also connects me to the land. My flock is hefted, not just in knowing where they belong but in being part of this place in everything from their genes to their gut biome, their feet adapted for our rocky hillside and fleeces adapted for shedding rain and snow. This intimacy between animal and place doesn't happen where animals are units of production, where the goal is nationally uniform herds of a handful of breeds that produce the most meat in the shortest amount of time. The things I strive for in my flock—like strong mothering and hardiness in a winter climate—are bred out of modern herds. Modern breeds are totally different in looks and behavior from the more primitive breeds like ours that were adapted to raise themselves and their young with little help. Modern sheep breeds are the equivalent of a supermarket tomato, bred to be generic. They can't thrive on what nature provides, on grass alone.

OVID IS CURRENTLY MY OLDEST RAM AT SIX YEARS. I CAN easily spot his long-legged daughters and granddaughters among the flock, and some have sold around the country. He has been the dominant ram for three years since coming into his prime and is gentle and slow, like an ox. Some of the young rams are skittish, but Ovid never breaks his air of stately command.

When the time comes to reintroduce rams to one another after the rut, it takes some thinking. A dominant ram with much younger, submissive companions will be friendly. Rams of the same age, in their prime, will almost always fight to figure out who's boss. They fight by backing up and running at full speed into a head-on collision, the force flinging their back ends up in the air. I usually pen them in a small safe space for a few days, small enough that they can't run at each other. I throw in a few hay bales or an old tire to make their movement even harder, and I give them plenty of food and water. Stress will make rams fight more. Isolate a ram at breeding time, and he will only beat himself up on the walls of the pen, trying to escape. They need to be together, and in a few days they have worked it out and are eating with noses an inch apart at the feeder.

Even after all these years, I make mistakes when I know better. I thought none of my less dominant rams would mess with Ovid, but now I see how wrong I was.

Ovid's long, refined nose is bloody and tattered. Blood from a deep cut between his horns has run down his forehead, parted into two rivers down his cheeks, and dried to a dark-red crust like stripes of war paint. Streaks of crimson spatter his horns and mane of white fleece. His left eye is swollen shut and the size of an orange.

The young ram lambs in his new field knew enough to stay out of his way. And, like a kind uncle, he scarcely deigned to discipline them. An older animal in the group of rams, Bronwen, was not a fighter. But Chadwick, three years old, was another story. If Ovid was past his fighting prime, Chadwick had just come into his, and when I brought Ovid back into his orbit yesterday after two months apart, Chadwick saw the chance to change the pecking order, to claim the succession of kings.

I purchased Chadwick from a friend who was going through a divorce and had to move and sell all her animals. I didn't need another ram, but I also don't like to pass up the chance for good new genetics when I find them. Having sold so many rams around my region for so many years, it's often hard to find one I can breed. I saw this ram's beauty right away: broad forehead and straight nose, thickly muscled shoulders and long legs, jet black. (I named him after actor Chadwick Boseman, who had died the day before I brought him home.)

Looking at him now, I notice how his horns are nearly twice as thick at the base as Ovid's and come together over his forehead like a solid shield, making a natural helmet that protects the bones over his eyes and the bridge of his nose. His head is both sword and shield. One tiny scrape on his brow and a crown of white fleece tangled in his horns is the only evidence of his dethroning the older ram.

As I watch, Chadwick mock charges, and Ovid drops his head and moves sideways, a gesture of surrender. Then, no longer able to stand from the trembling, he crumples down. The other rams eat, stepping around him. A small white ram nuzzles Ovid's face for a minute and cleans a little of his blood like a mother dog.

I go to the house for a bucket of warm water with molasses in it, a home remedy the equivalent of my mother making me hot lemon and honey tea whenever I had a cold as a child. When I heave the bucket over the fence, the other rams aren't interested, but Ovid gets unsteadily to his feet and drinks in great gulps, slurping it like a horse after a hard run. He still looks shipwrecked, but the trembling has stopped.

Ovid lies down again in snow darkened with his bleeding.

Bronwen limps over and stands above him in a protective way. Bronwen is neither ram nor ewe but something of both. She's very old now and has arthritis in her left foot. She stares at me with those iridescent gold eyes and rectangular pupils that all sheep have, inscrutable. She arches her neck and scratches her back with one long horn.

When Bronwen was born as one of triplets, she looked like a ewe lamb. She was black, with swirls of gray over her fleece. But soon she became a shape-shifter. Her black face turned gold, and her slender horns grew thicker at the base and very long and delicate at the tips, curling out more elegantly than any sheep I'd ever seen. From a tiny runt she grew tall, with narrow hips and beefy shoulders. Around the ewes that fall she began to curl back her lips and stroke their flanks, behaving more like a ram. She didn't conceive her own lambs. The rams seem to perceive her as a low-level threat. They will keep the ewes in heat away from her, but they won't outright fight her like they will each other. Now the oldest animal in the flock, Bronwen—who is an intersex animal, also called a freemartin—is still hands down the most magnificent looking. She has a silver body, a face that is gold in winter and black in summer, and horns like something you'd see in the Serengeti. Look up images of a wild mouflon, and that is what she looks like. When she sleeps, she lies on her side and rests the tip of one horn on the ground as you might prop a bike on its handlebars, her head suspended in the air.

Bronwen has always had a kind of magic to her. She won't let me touch her; she's aloof and difficult to handle at shearing time, but I'm more loyal to her than to any sheep I've ever had. When I'm moving the sheep in summer, she is the one who knows the way, and the others follow. She sees openings that the others don't see, as if she reads my intentions. As the elder, she carries the most memory of the land and is hefted to each place where the flock spends time. I've kept Bronwen all these years because I was told to by a Navajo shepherd who gave her a special blessing. Bronwen's blessing—and all that I learned from the friend who gave it to her—is related to this idea of a flock or a breed coevolving with the land and with the shepherds who live on that land. It's also related to the patterns of damage and healing that play out on the land in small spans of time and large ones, and it begs a question that is always present for me: How can we keep trying to understand enough to promote healing and do less damage?

DINÉ (NAVAJO) SHEPHERD AND WEAVER ROY KADY CAME from Arizona to a leadership gathering at our farm in 2010. Kady (I'll call him this so as not to confuse him with my vet, Roy) stayed a week with us, and he had an instant connection to our Icelandic sheep, who resemble the Navajo-Churro with their horns, their short tails, and their dual-coated fleece. About 1,400 years ago, I learned, they shared a common ancestor in northern Europe, but the sheep that traveled to Iceland with the Vikings evolved to live in the cold, while those that went south into Spain evolved to live in the heat.

It was a hot day ten years ago when Kady walked with us down to where our flock rested and stamped at flies under the wild apples. He immediately singled out Bronwen as the leader. He burned some sacred tobacco under the hum and flicker of the trees and sang songs to her in Diné. To his people, he told me, an intersex sheep was sacred and brought luck to the flock. "That sheep, there," he said, pointing to Bronwen's bright-gold face with wide-set eyes and flaring horns, "should never be slaughtered."

Kady was wearing a T-shirt that said "Sheep is Life," from the tribal gathering he was part of each year to bless the sheep, share creation stories, and come together around weaving and traditional lifeways. In his soft voice he said to me, "Sheep is your backbone. It's your lifeline." And then he began to tell me part of the story in which his people and sheep came together into this world to sustain one another, as gifts to one another, and how his people's sheep had almost been lost. Kady is a weaver, a tradition passed down from his mother, and he talked about the rugs he weaves, and the loom itself, as sacred expressions of the union between earth and sky and sheep and life force. I couldn't begin to understand it all, but it made a deep impression on me. It led me to dig deeply into the story of how the first sheep came to North America, and how they—and essential traditions of the people who were culturally, spiritually, and economically entwined with them—almost disappeared.

English colonists brought few sheep with them and did not know of the native wild sheep in the West. The first domesticated sheep in North America belonged to the Native tribes of the American Southwest as early as the 1500s. According to the Spanish, in 1598, Juan de Oñate brought Churra sheep from the south to pueblos along the Rio Grande and Rio Chama in what is now called New Mexico. The Spaniards chose to load their ships with Churra sheep (also called Churro), not the more highly refined Merino, because of the Churra's hardiness, long legs for walking long distances, and coarse fleece that would shed if not sheared. Years later, Spanish soldiers reported large flocks of Churras being moved around the mesas of what they called Navajoland, which the Diné people themselves call *Diné Bikéyah*, "The Glittering World," their homeland centered between four sacred mountains.

According to the Diné, however, sheep and goats did not arrive with the Spanish but are part of their creation story, given to the people as a life-sustaining gift by the *Diyin Dine'é*, the holy spirits who also control the wind and the rain. The Blessingway is the ritual enactment of this knowing that the lives of the Diné people and sheep have been entwined since the beginning of time. The annual ritual blesses the flocks, gives thanks to the spirits, and preserves the knowledge for the people.

The Blessingway is more than a wish or prayer for good health and more than a gesture of gratitude. The Diné perform it as an embodiment of the creation story in the present moment in order to restore and protect the sacred balance of people's original relationship to Mother Earth. This balance with nature that is maintained through spiritual connection is something they call *hózhó*. Hózhó is not static or passive but requires continuous presence and right relationship.

White settler encroachment into the Diné's homeland dealt several violent blows to their sacred hózhó. First, in the years during and just after the Civil War, the US government used the excuse of "securing the Western frontier" to force Diné families to leave their homes and flocks and walk hundreds of miles to Fort Sumner. At Hwéeldi, as the Diné called the prison camp, many died in appalling conditions of starvation, disease, and despair. An underlying reason for the removal was to punish the Diné and prevent them from raiding stock from white ranchers who were constantly pushing into Navajoland and competing for grazing land.

During the first two decades of the twentieth century, a period of homecoming from Hwéeldi and of good rains, Diné flocks and families grew, and yet the traditional practice of transhumance—of moving their flocks in a seasonal pattern language mapped to their knowledge of the land—was disrupted by being squeezed into an ever-diminishing and regulated space. This forced them to continuously graze the same areas and visit the same water sources, and it set the stage for the rapid degradation of their land as the rains ended.

The 1930s were a period of severe drought. This, combined with more animals on the land, led to widespread overgrazing and erosion throughout the West, though perhaps nowhere was the problem more studied and controlled than within the Navajo reservation. Although the government had encouraged the Diné to grow their flocks all through the 1920s as an economic recovery effort, by 1930

Bureau of Indian Affairs (BIA) agents were alarmed at the condition of the land and decided there were too many animals on the range.

The flock-reduction years, starting in the 1930s and building in intensity until 1950, were devastating years for the Diné. As well documented in the deeply researched ethnography by Marsha Weisiger, *Dreaming of Sheep in Navajo Country*, United States government agents slaughtered sheep and goats by the thousands, insensitive to the fact that the Diné depended on their flocks to feed their families with milk and meat, to establish social currency, and to provide real income in the form of wool that women would weave into rugs and sell. More than that, they ignored that the Diné regarded their animals as a sacred gift, and the slaughter of them to be an act of disrespect to the spirits that controlled the rain.

The BIA agents and range scientists who led the effort to cut down the number of animals grazing the mesas knew that their tough measures would impoverish the people. Their misguided solution to this enforced impoverishment was to try to "improve" the Navajo's sheep by setting up stations to breed female Navajo-Churro to Merino rams, hoping to convince the Diné to keep fewer sheep but a different breed, presumably of greater commercial value. But the Merinos' finely crimped, greasy wool—which brought top dollar in commercial markets—was useless to the Diné women weavers, and the sheep themselves were less well adapted to the terrain. What the rangeland scientists knew, but chose to ignore, is that all traditional pastoralists around the world practice subsistence, meaning that they keep their animals for multiple reasons that have to do with bonding, family ties, and cultural rites as well as meat and milk and wool. These sacred animals have coevolved with their people and are specifically adapted to their lifeways and needs. To Diné shepherds, a sheep was not just a sheep, any ram not just a ram. To be told they must now breed what was left of their decimated flocks to a Merino ram must have been worse than insulting—another violent betrayal. The Merino was the colonial ideal of a sheep; to the Navajo shepherds, the Merino rams breeding with Navajo-Churro ewes represented the end of their flock, not the beginning—an enforced loss of cultural identity as strong as the enforced loss of language.

The breeding program was a complete failure. By the 1950s, their *Dibé dits'ozí*, the sheep that the Diné had bred for over four hundred years to adapt to their land and to produce the perfect wool for their rugs, had nearly died out. (A few flocks, hidden away in remote canyons, held on as refugia, seedbeds for later revival. Kady and other shepherds in Arizona and New Mexico are protecting the Navajo-Churro breed now and using their wool again for weaving.) What's more, the land also failed to recover during and after the stock-reduction years; invasive species like cheatgrass and thistle dominated vast areas of the range. Navajo land patterns were disrupted by the regulations and boundaries forced on them. Weisiger explains in her ethnography how, as traditional livestock husbandry broke down, overgrazing actually intensified.

Traditional ecological knowledge is a phrase used to describe the generations of passed-down wisdom pastoralists hold about their flocks and landscapes and how both are interwoven with their survival as people. The Diné's story—in which their traditional ecological knowledge was utterly disregarded by the range specialists who insisted they were helping them protect their land—has been repeated and is still repeating for traditional pastoral people all over the globe. William Cronon writes in the foreword to *Dreaming of Sheep in Navajo Country*: "That these consequences were unanticipated and came as a surprise to the New Dealers . . . demonstrates that it never occurred to them to 'think with sheep.' For these managers, sheep were just unwitting biological agents of the ecological destruction of desert grasslands. They never recognized the extent to which Navajos and sheep were part of an integrated cultural universe in which the destruction of one could hardly help but have tragic consequences for the other."

Meeting Roy Kady and being inspired by his visit—and by my friendship with two Diné women who came to our farm later on—all made a deep impression on me. I do not carry their ancestral and cultural wisdom, but I now have a sense of how our life on the land requires a cosmology in order to embody health and balance. The approach guided by a worldview of separate and inanimate parts, which is the system of farming taught and practiced in our dominant culture, fails us over and over again. This much is clear. To make one's own cosmology is possible, I believe, with enough attention and attending, and that cosmology becomes a guide for how to respond to change.

OVID'S EYE IS EVEN MORE SWOLLEN THE DAY AFTER THE FIGHT, oozing blood and yellow puss. I carry down three green metal panels, each six feet long, and make a small pen against the gate of the field to trap him. He is as slow as an old man with a walker. He bends his head and rests it on my leg, and I take off my glove to reach my fingers down through six inches of fleece to his skin to give him a scratch. He's being more docile than usual, which isn't a good sign.

I hear the vet's truck on the snowy road to the barn, his door and then tailgate slam shut. Roy comes over the rise with his black bag and brisk walk. "Whatcha got?" he shouts to me and grins a hello, but he immediately turns to the ram as he steps through the gate.

"Old man, you got yourself into some trouble, didn't you? Don't you know better than that? Who was the villain? Oh, yeah, I see him. That was no contest."

He keeps up the running conversation with Ovid as he probes the ram's neck and face, sticks a thermometer in his rectum, palpates his abdomen.

I stand at Ovid's head, holding him still, but he is in a stupor. His breath comes in white gusts from his nostrils.

Roy sticks Ovid with two needles: antibiotics, and Baytril for pain. Then he pulls a scalpel out of his kit and opens a long slit next to the crusted-over wound in the top of Ovid's forehead to let the hematoma drain and let up the pressure behind the eye. Blood showers onto my boots and blooms scarlet in the snow.

"He looks like Rocky," Roy says. "All that blood flowing in the eye socket will make it grotesquely swollen. It might be a while before we can tell if the eye itself is intact."

Roy says to Ovid: "Right! You can take it from here, my friend." Then to me as he leaps over the fence and is striding away, he adds: "That guy should really be in the barn staying warm."

I'm still mad at myself for making this mistake. In the early days I didn't know better, and Roy also came to my rescue then. The first ram he ever patched up was Osprey, who was born our second season. Osprey was so huge at birth that he almost tore apart his little mother, Kali, and by the time I finally delivered his massive head with its huge pointed horn buds, both Peter and I were crying. Osprey was piebald, like a Pinto horse. He was the first ram I fell in love with and kept for breeding of my own. The next breeding season he and another ram smashed a hole through the fence in the night to get at each other. I went out in the morning to find Osprey's lower jaw hanging from a hinge of skin. Roy came, and we laid Osprey down in the snow, sedated. I helped Roy bolt the jaw back together with a large piece of metal, like something you'd use to hitch a plow. Roy warned me it might not hold, and it didn't. When Osprey could no longer chew a few months later, we shot him to end his suffering. I told myself I'd never let my rams kill each other again. That was fifteen years ago, and now here I am. I'm not sure Ovid will make it.

I loop a rope halter around Ovid's horns. "Here we go again, buddy," I say, as the insult of the halter touches his naked wounds and awakens an inner rage. I shove the gate open against the snow and try to hang on to the rope as Ovid staggers back up the long shoveled path to the barn, ornery with pain. He bucks and leaps, alternatively charging me and lunging forward, pulling me off my feet. He outweighs me by at least fifty pounds.

"Taking the bull by the horns," should be an expression of foolishness, not bravery. Somehow I get him in a pen in the dark barn with only one sharp whack to my shin.

"All right, old man," I say. "I guess I deserve that after what I put you through."

For two days Ovid doesn't eat, and I can hardly look at his face. I give him salt and kelp to tempt his appetite and watch him lick it off his lips. He hangs his head still as I press a hot flannel over his eye, drawing out the pus. On the third day, the socket drains colors of purple, red, and yellow, and on the fourth I see a glint of something that might possibly turn into an eye shining out of its swollen lid. The eye emerges from the depths, watery and solid white, like a reflection of the moon.

It's rare for any sheep to tolerate being alone, but Ovid seems content in his convalescence, giving me a throaty hello when I walk into the barn to flake out hay for him. There is a gift in my mistake, and in this restoration. Caring for him, I think a great deal about healing. All that is wounded in nature, given enough time, has the impulse and imperative to heal. To heal means to reknit, to bring parts that have been separated and damaged back together, to become some version of whole again. Along the way, though, the healing process always changes something. I am beginning to understand that healing is not about returning to what was, but about accepting the change and adapting to the brokenness. This is happening all around us, for people, for the land. People have done damage to the earth and to each other that can't be undone. We can lament what was, but that won't help us take care of what we still have. In fact, it might just hold us back. Nature herself keeps giving and never giving up. Can we be like the trees that keep growing to seal over barbed wire, like the Diné who hung on to their ceremonies and traditions and are reviving their Navajo-Churro flocks, like the injured ram who will learn to see danger coming with one excellent eye?

INTERLUDE III

To love is to join.

—JEAN GIONO, *THE SERPENT OF STARS*

I PARK THE TRUCK, WALK TO THE LOBBY AND HIT THE BUZZER, put on a mask. The pandemic has mostly passed, but the young man at the front desk still needs to check my temperature by holding a small plastic gun to my forehead before letting me walk up to my mother's room on the second floor.

A nurse is giving her a shower. Mom is sitting naked on a stool in the shower stall, and the nurse, Dee, is standing just outside the curtain, spraying water with a handheld nozzle and rubbing Mom's back. Mom's body is as small as it ever was, her tummy flat, her breasts folded down like pockets in a coat. Her skin is mottled with brown spots and little moles but is surprisingly firm-looking, not hanging like lacy drapes as one might expect at age eighty-four.

Her lower legs and feet, though, are deep purple. Her capillaries spill beneath the skin like dark-red ink on old parchment.

Mom looks up when she hears my voice, peers through the water, sleep grains caught in her lashes, hair plastered to her head. She squints, distant, uncomprehending, her smile more of a question than an answer.

Another nurse comes into the small bathroom, and together the two women gently but firmly convince Mom to let them lift her weight off the stool, turn ever so slightly, and sit down again in a wheelchair. She is terrified to stand, sucks her breath in, clutches hard to the shower rail and won't let go until, finally, I have to pry her hand away.

Instructions like "Let go with your hand," or "Step back with your foot," no longer compute; her brain cannot make sense of that in physical space. So any moving we do scares her—it is all a surprise and a potential toppling.

She has lost words, and with them the ability to walk.

Dee wheels Mom into her bedroom, and I put Mom's pajamas on. I give her a very long hug, her wet head against my belly.

"I love you, Mom. I'm sorry I haven't been here in a while, but I didn't want to give you my cold. You know I'm your daughter, right? I'm Helen. I love you so much."

Often if she doesn't remember me, I will show her photos of the farm and of the lambs, especially one photo of a few years ago when she could still get out to the barn. She's sitting on the steps with Chance and Lightning climbing on her, and she holds a pink baby bottle, her face beaming. The lambs always trigger her memory of who I am and where I live.

She looks at me again, and suddenly, for a moment, she is all there—her utterly loving and present regard.

"Yes, darling. My Heli. And I love *you*," she says.

In that moment I feel her entwining, her regard like a strong tendril, holding me up. I feel Wren in that moment too, and the tenacious vine that grows and holds us all—mother, daughter, mother, daughter—even as the ground crumbles beneath us, as we lose time and space and words and even each other.

Dee rubs Mom's legs with cream, and under the pressure of her hands Mom's feet flash white, then purple again. It looks as if she's been found on Mount Everest barely alive, like her legs have been frozen solid and she might lose her feet. The purple is so much darker and more widespread than last time I saw them.

"She's just one of those who get purple feet," Dee says, as if reading my concern. "We all rub her. Everyone loves Ruthie, so she gets spoiled."

I imagine Mom being rescued from a glacier, how miraculous it is she's been found alive. How her mind is still out there, snow-blinded, altitude-blasted, meandering the edge of a precipice in the alpine zone forever, part of the clouds and the sky and the realm of ravens and snow leopards, of monks and angels.

THE NURSES LEAVE TO GIVE US TIME TO VISIT. I PUSH MOM's chair toward the French doors and the little balcony looking out on the back road. She hasn't had a haircut for a long time, and something about the wisps of white curling around her neck make her look very vulnerable. She always had short reddish-brown curls that she brushed with her fingers.

I pull *The Serpent of Stars* from my bag. My life in books has returned, and I am starting to write again. It is hard for me to describe my love of this very strange, very small book that I found in the Tempest Book Shop so many years ago, Wren an infant strapped to my chest while I was looking for bread. It has become a talisman. It lives on my desk and is scuffed and underlined, wavy with use. Its compressed, sinuous imagery has given me a timeless lens through which to see the mundane, repeated rituals of my life on the land.

I have always read to my mother—in the years she lived in the farmhouse alone I'd visit and sit on the kitchen counter reading aloud while she made dinner. She loved the Brontës, Jane Austen, George Eliot. She was a quick cook and made simple meals. Carrots and sprouts and sweet potatoes braised in oil and piled onto a tortilla with fresh tomatoes and herbs, cheddar melting over it until it sizzled into a crisp lattice on the pan.

There was never enough time; the pages I read were a crystal of salt on the tongue.

Sitting in front of her wheelchair, knees to knees, I read a few pages to her now. Whenever I look up, she has her eyes riveted on me. I read with my full body as if I'm on a stage for her, raising my arms, gesturing with my hands, conveying the strange images in a language older than words.

I am reading a chapter in which the land is described as the sea, the sheep as water.

> Because the occupation of the masters of beasts is something like water which runs through the fingers and which cannot be held. Because that odor of suint and wool, that odor of man cooked in his own sweat, that odor of ram and goat, that odor of milk and of full ewes, that odor of nascent lambs rolled in their slime, that odor of dead beasts, that odor of herds in the high mountain summer pastures, that is life, like the brine of the great seas.

She laughed at *man cooked in his own sweat*. She keeps nodding her head when I ask her if she wants to hear more. I begin to think she understands it better than most, though perhaps my wild gesticulations help. (If you are ever reading to people with advanced dementia, I recommend Jean Giono. He writes from a place where logic, or any effort at full comprehension, only gets in the way.) I continue:

Four huge fires lit up and defined the large stage of grass and earth.

Right in the middle, a man was standing. He was waiting for what would flow from his heart. I remember that he was a tall, thin man, one who saw things, who feasted on visions. His nose turned into a bird's beak under the fire's high flames. He was wearing a red scarf on his head, tied gypsy-fashion.

Suddenly, he raised his hand to greet the night. A rumbling flowed from the aeolian harps. The muffled flutes sang like springs.

"The worlds," said the man, "were in the god's net like tuna in the madrague . . ."

You could have heard him on the other sides of the earth and the sky.

The chapter finishes here, and I put down the book. I look up at her, sitting in her chair, listing slightly to the left. Our knees touch.

Mom says, "Wow! That, of all things gets . . . all the cracking and cooping . . . where I've been . . . like . . . umph . . ." and she thrusts both arms and her lower jaw out in a gesture of defiance.

"Exactly. You got it, Mom." We laugh together, about all of it, whatever all may be.

THE BOOK IS AN ENIGMA TO ME STILL, BOTH SO SPECIFIC IN ITS squeamish detail and mythic at the same time. But it gives me a feeling. It describes to me how, through a relationship to the flock, everything in the natural world can become more animate, more internal. The boundaries of self and world *can be* erased. Shepherding can be what I do in the world, or it can be what I *am*.

No moment describes this better than the scene in which a lead shepherd, Bouscarle, comes to a young boy learning the trade and lays a sweaty hand on his shirt, his eyes gleaming: "'My boy,' he said, 'don't think you know everything. You know the sheep, but to know is to be separate from. Now try to love; to love is to join. Then, you will be a shepherd.'"

To know is to be separate from. To love is to join. How many times have I felt I couldn't act or be worthy or seen until I really *knew* something, thinking that my love was not enough?

I kiss my mother's head goodbye, kiss her small dried leaf of a body with her purple feet, and I kiss her fading mind. Most of all, I kiss her love, which is infinite and unconditional and has always been and always will be. *To love is to join.* She taught me this; I see it now as everything else falls away.

CHAPTER 12

WINTER

> The beasts sighed, all together. The hills fell silent. The man gave a voice to the joy and the sadness of the world.
>
> —JEAN GIONO, *THE SERPENT OF STARS*

THE LONG WINTER IN NORTHERN LATITUDES HAS THE weight of lead; it seeps into my bones and makes my tread heavy over the fields. Some days, when the clouds hang low over the mountains and the semifrozen river breathes wraiths of fog from below, it never gets light. A steel lid is clamped over my head. I am perpetually half-awake.

Nothing stirs. My whole half-lit world is waiting for something, suspended. Waiting for a penetrating wind or a blizzard, for a soft snow that falls thickly onto the voided fields. Waiting for a thaw, waiting for a rare flock of snow buntings or a common crow grumbling at an owl, or waiting for nothing at all, just waiting.

My blood congeals like frogs in torpor below the mud in the pond. My mood dips low and will emerge only as the light returns. I am kin to snake, to sunflower, a creature of warmth and light like any other. But I have become grateful for this season of integration. I enter deep time, my mind emptied of the endless to-do lists of summer, of people to gather and deadlines to make in the short intense harvest season, the season of grass and berries, of proliferating insects and weeds and customer demands. In winter, when the farm is quiet, when everything freezes and congeals, migrates and gestates, my heart opens to longing.

Longing is a feeling that rises when I have time alone and have put down my ambitions; it leaks out of what I thought I had contained, blooms into empty space. My mind disperses, like a glacial river moving against the ice deep underground, collecting, moving to somewhere old—to my lineage of women and how they endured, what they left me to endure— and to somewhere new, revealing what is hidden.

I used to despair at these times of melancholy that have come in the dark months all my life, but now I understand my mind and body in a seasonal way. Winter is slower, more contemplative, and even the strain of sadness has a strange sweetness that connects me to insights and memories I've been missing all the other seasons when I move too fast. Winter is when I sit down. I begin to write again. At a table by a window each morning, I look out on the frozen farm pond and meander through the till of words.

THE STORM BEGINS SOFTLY, SILENTLY. NOT LIKE RAIN shimmying down the roof shingles or slapping against the windowpanes. If I didn't look up from my reading and press my face against the frosted glass of the old farmhouse, I wouldn't know, but there it is, pummeling out of the darkness, one of the most magical of all nature's phenomena—countless zillions of iridescent crystals swirling and falling and drifting into an unblemished carpet over every maple twig and fir needle and rooftop and stone wall, over the sheep themselves lying like mounded glacial erratics in the field.

The storm continues all night and into the next day. We wake to a profound silence. Only the wind stirs, far away on the hillside. I dress in the cold, gritty mud room where we store heaps of old clothes amid bins of potatoes, garlic, winter squash, and onions in various stages of imperfection. I pull on a down vest, quilted coveralls, two pairs of socks, a hat, gloves, rubber boots, and a strong canvas coat. Peter and I shovel paths from the house to the barn, from the house to the driveway, from the house to the woodshed, and finally a path for the sheep to follow out to the water tank. The sheep will lie in the drifts, but when the snow gets deep they won't walk through it. Some ancient instinct tells them they will be too easily buried, their thick fleece acting like a plow that heaps up more snow than their frail legs can push against.

The sheep lie in the field as the snow pounds down. Snow over their eyes, covering their noses, piling up on their heads between their horns. As the snow builds over their backs, their thick, greasy, almost impenetrable layer of fleece insulates them from the cold and wet. Their hardiness is legendary—I have read stories of sheep being drifted in place in Iceland, completely buried under the snow, sometimes surviving for many days in a pocket of breath underneath. But on wind-drifted plains where ranchers can't get food to them, snow is a killer of sheep.

I stop to rest from my shoveling. The sheep seem impervious to the storm. I imagine they feel like I did as a child in my grandmother's bed when we went to England to visit. Some instinct made her fiercely tuck me and my sister into bed with so many layers of taut, thin wool blankets that our small bodies were pinned. We had to lie on our backs until the next morning when she would return to liberate us, throwing back the heavy window drapes and our confining bedcovers with a commanding gesture; dawn held promise.

This was a gift my grandmother gave me: to welcome dawn. No matter how early I tried to wake up, she and my mother and grandfather had already done so much without me—collecting the glass milk bottles in their metal basket left on the doorstep, hanging out the wash in the concrete yard between the flats, or walking into the village for the bread. My grandparents' tiny English garden was a universe in microcosm where my sister and I would crouch to find treasures under leaves and hedges: dead birds, worms, cigarette butts, once a spiky hedgehog like the ones we'd seen only in English picture books and didn't believe were real. Gran taught us about broad beans, pollinated by hummingbirds, rough and enormous but tender when picked at the right time. She showed us how to layer the kitchen scraps over the dormant raised beds and then to layer newspaper and grass clippings, how to carry out our bathwater to water the potted plants and beds of strawberries. Everything was small and tidy and had a purpose that we were required to learn. Within her miniature universe, my grandmother's presence was humble but huge, always moving, and we wanted to be in her light. I felt deferential, but I also knew she liked naughtiness; anything that went askew would make her laugh, and any time she laughed my grandfather would smile. He was a quiet man, a great gardener and reader, who walked fast and talked slow. He adored my grandmother.

My grandmother died in her sleep one month shy of her one hundredth birthday. My mother, in her seventies then, was flying home from my grandmother's funeral when she felt unwell. On her way from Boston back home to Vermont, she drove straight to the emergency room, where she ended up having abdominal surgery and then spending two weeks in the cardiac wing to settle her fluttering heart. Shortly afterward, my mother's mind began to place itself elsewhere and lose the thread of this world.

As I walk through the snow with a sledload of hay to feed the rams, spinning flakes stinging my cheeks, my grandmother is with me, and my mother too—pragmatic English women who were rarely kept inside for any reason or by any form of weather. I would give anything to walk fast with my mother again, or even slow, but she was always most at home at a brisk pace, in the wind. Memories of my grandmother visit me most often in the garden, or when I'm planting seeds. She used to get impatient and dig up the seeds she'd planted in pots on the kitchen windowsill, "just to see if there was a sprout."

One by one the mounds at my feet rise and shake vigorously like dogs after a swim. Slowly, a flock of sheep emerges out of the field, walking toward me with frosted muzzles and hoary eyes, bodies steaming. In the silent, cold air I have a marathoner's sense of pacing—slow, in for the long haul. I take in the silence, the wisps of wind across the white fields, the ice crystals on the sheep's whiskers. The wind comes straight from Canada, smelling of resin and iron, making my hands ache and crack. I blow into my stiff gloves and slap them against my thighs as I walk. In the house, I strip down, sit in the beautiful rocking chair that Peter gave me to nurse Wren when she was born, and hold my hands and feet out to the fire until I can feel the painful throb of blood returning.

Even in the cold, I don't begrudge the chores. The daily necessity of bringing food to the animals and watching their pregnancy grow week by week, anticipating new life, gives me an anchor to joy, frail but certain. The lambs will come on schedule even if spring takes her sweet time.

FEBRUARY IS "THE POOR MANS PICK-PURSE, AND THE MISERS cut-throat, the enemy to pleasure, and the time of patience," says the *Kalendar of Shepherds*, one of the first books to depict the seasonal ritual of the farming year, first published in 1493. Woodcuts for each month of the calendar show men and women doing the work of the land. For reasons utterly unclear but also delightful to me, this book is also referred to in literature as "The Compost."

Each month of the Kalendar begins with a prose poem, describing the time of year. September, when "the windes begin to knocke the Apples heads together on the trees, and the fallings are gathered to fill the Pyes for the Houshold" holds "Winters forewarning, and Summers farewell." Winter comes, and December is "the purgatory of Reason." In January, the shepherd "hath a bleake seat on the Mountaine, the Blackbird leave not a berry on the thorne, and the garden is turned up for her roots."

The Kalendar, for the Middle Ages peasant, was a collection of stories, a charting of the seasonal year to offer perspective and hope. In northern cultures of the globe, people have relied on telling their winter stories to get them through. Elemental stories. Stories told and repeated and passed on around the fire, the kind people tell when the nights are long and there seems no point to the darkness. Stories of hardship and effort, and most of all of the continual cycle of things. The storms and darkness and light that shift their shapes over us again and again are crafted by the tellers of stories into both consolation and remedy. Stories are like the salt stones the sheep wear smooth with their lips, seeking nourishment.

These generations of sheep that are conceived in the dark months and born as the light returns create my Kalendar. They are the line that cuts through my years. This ritual of breeding and birthing, of work and rest, happens year after year—even as the climate unravels, the storms are more frequent and violent, the winter less predictable, the seasons without steadiness. There is a constant, seasonal farming rhythm that must go on and that is anchored by the blood and bone, the milk and muck of these animals, despite the temperature of the day or the shifts in the seasonal pattern. Despite the barometer of my mind. Everything shifts, and I go on. Everything changes, but the lambs still come and the stories pile up with the snow.

MY MOTHER HAS LOST HER STORIES. SOMETIMES I SEE HER searching for them, see a flicker of panic behind her usually peaceful eyes. She tilts her head, an almost imperceptible movement of resignation. All my life I've counted on my mother to have words—soothing words, wise words, somehow always the words I needed to hear. She was the one who was there when life got me down, or when I simply needed to hear my thoughts out loud. I wonder how to live more peacefully with my longing for everything about her, most of all for the deep conversations that she and I shared. It can feel like nothing but loss.

I wonder what it feels like to her. In leaving her home and coming to this place, knowing her mind and memory were going, she was facing becoming lost without the possibility of return. Is "home" for her now a beloved's hand in hers? Is a slight squeeze, or a smile, a trace of recognition of something akin to the feeling of having been gone and having returned?

Now our relationship is on the physical and spiritual plane, a kind of simple being-togetherness with touch and words that console but don't require response. In my intense longing to retrieve her, I can feel my heart reaching out and touching her, and I feel hers touch me back, even when she doesn't in any conventional way know who I am. I sense she knows *what* I am—someone who holds our love—even if the *who* is lost. She holds a fragment of me still.

AFTER THE BLIZZARD, THE WIND HAS PICKED UP AND THE temperature drops by the half hour. The ewes pull tufts of hay from the feeders with vigor. After a bitter night, they need to replenish their inner fire and feed the lambs they are carrying.

As I lay out another armload of hay into the wooden racks along the eastern wall, the ewes crowd me to get the first taste, pinning me in on all sides. Their wide pregnant bellies make a solid wall of soft flesh and wool between me and the rest of the shed, and I push my way through, like wading through waist-deep water.

A red hen balances on a big silver ewe's broad back, swaying, crooning, not willing to let go of her warm perch. More hens perch on the hay feeders and on the swinging door that separates the barn from the sheep shed. They have abandoned their coop this winter in favor of this warmer space that is alive with sheep droppings and hay seeds.

The ewes sniff and jostle, seeking the best tufts of hay and hoarding their stash from the others with a swing of horns. Velvet-lipped Habibi stands behind the sheep and cranes his long neck over them toward the food, above it all. To us the bales mostly look the same, thick slabs of dried grasses, flowers and leaves of dandelion and buttercup laid down by the mower, dried in the sun and baled last August to store until this cold snowy day. But to the sheep each bale has a different taste and smell; each is from a different field or part of a field, finer grasses from the shady edge, broad-leaved plants and clover from the richer soils with more sun. Each bale is like a different menu to them, and some bales they sniff and then look at me with longing eyes, asking me if I have anything better that day. I need more than this salad, they say. I want the red clover like a richly spiced stew, full of protein for my growing lambs.

I try to find them the best bales, holding the flakes up to my nose. I have learned to tell what will taste the best, what bears the sweetest, most nourishing leaves of grass.

I, TOO, LIKE MY MOTHER, AM LOOKING FOR WORDS, LOOKING to make sense of fragments. Unexpectedly, my story of the land—and of the twenty-five generations of sheep who have taught me to observe and to endure—is also about being a daughter to a mother who is dying and a mother to a daughter who is leaving home. In the long process of saying goodbye to beloveds I feel a sense of rupture and yet also a greater awareness than ever before of the need to make meaning, to write things down, to hold what is. I have come to a profound juncture. I don't want loss to upend me. I am calling on things I have learned in this shepherding life of grace and joy, of death and risk, of endings and beginnings, of edges and the unknown, of letting go. I am searching for those stories, stories like salt, stories about mothers and grandmothers, about daughters.

I found one such story, with origins as long ago as the *Kalendar of Shepherds*. In pastoral regions of Scandinavia, it was exclusively women's work to milk the livestock and make cheese, so when spring came, the women and girls would leave the village to spend all summer in alpine pastures. To make a go of it in the far north, pastoralists had to follow the summer grasses vertically up the mountain slopes, above the shade of trees and stone. The warmest pastures were halfway up the peaks, a long way from the shelter of the winter homestead.

High in the summer farms, in simple shielings made of piled stone with a sod roof, women prepared wheels of cheese and butter to carry back with them when the days turned short and cold. This was the stored protein and fat produced from mountain meadows by the animals, which would sustain the village all winter. Surplus was sold at weekend markets, with the men coming up on horseback the night before market day to collect the cheese and consort with their womenfolk in the long summer night. To celebrate the coming of the menfolk, the women would make festive food like woodsy *setermat* from buttery whey, smoked salt meat, and wild berries, or *ystingsoll* out of cream, fresh curds, and prim. This cultural practice of *fäbod*, dating back to the Middle Ages or earlier, reached its peak in the nineteenth century in large parts of Sweden and Norway, but it is uncommon now.

The most alluring aspect of this cultural practice, to me, was the women's tradition of singing to their animals—called *kulning*—both to tell them where to go and also to protect them from danger. This elaborate ritual formed a tight bond between the woman and her animals, between the women who needed to work together across vast distances as they herded, and also between women and predators such as mountain lions and wolves.

This kulning music can be soothing and rapturous—high, long wavering notes and trills, often rising to float and linger. Many have heard strains of it in Disney's *Frozen II* when Elsa is called out of bed into the night by an invisible siren calling four mysterious notes over and over. She sings back, wondering if she should answer this call "into the unknown." Kulning can sound simple and haunting, as in that case, or highly ornamented and playful, or full of love and pathos. It can also be downright terrifying. Sometimes, sung with a ferocity needed to scare away a bear or mountain lion—or supernatural denizens of the forest—it is so shrill and piercing as to seem like the force of the Furies. Its language conjures black winged things with claws coming out of the darkness. Which was the point. It is a singing that was not meant to be performed anywhere but in the vastness of the alpine realm. I listen to a recording of women kulning and imagine how much more spine-tingling it would be to hear that vocal music in a thin mountain meadow, echoing off a wall of fog and stone.

Anna Ivarsdotter, a Swedish musicologist, defines this female musical tradition as a soundscape that forms its own ecology between women, their domestic animals, and the wild realm in which they graze. "The musical structure... consists of phrases of varying length freely combined into chains, the length of which reflect the demands of the actual herding situation. Since the grazing areas are vast, the music has to be heard over extremely long distances. And so it is. The vocal style is singular, totally different from any other type of Scandinavian folk music. This song as well as the sound of the horns and lurs can be heard up to three/four kilometers through the deep forests, echoing between the mountain slopes."

Women invented kulning out of their own longings, free of the village for the summer months, walking over vast distances with other women and with female animals and their young. Charged with protecting and guiding their animals, their wandering and calling had a purpose—to bring the animals safely home after the journey. The animals knew the voice of their shepherdess, and surely they felt its power. Shepherdesses had different vocal ranges for goats and sheep than they did for cattle—higher, more insistent. These small ruminants had quicker movements than the cows, and so the singing notes were also quicker to get the sheep and goats' attention. The tone would soften when coaxing the cattle to settle to be milked, and intensify when sensing danger. When looking for lost animals over great distances, women would use the songs to tell each other where they were, if they were safe, if the herd was together or separate.

"Usually two women would go together, the most experienced or the one with the most beautiful voice going ahead of the animals, calling them, while the other would drive them from behind, keeping them together. The task of driving was often given to the young girls and was excellent training for them. By listening and imitating, they learned the music, the special vocal technique and the use of the instruments. In this way the music was handed down from one generation to the other over the centuries," writes Ivarsdotter. Like the ewes passing knowledge of the mountain grasses to their lambs, so, too, the women and girls were being hefted—belonging to a place.

That such animal-human connection exists, and is one of the oldest forms of communication, seems completely logical. To have animals understand us would have been a matter of survival, and it likely also developed our sense of compassion and reciprocity for other living things. We heard their song, and we responded. They heard our song, and they responded. It's reassuring to me to know that our time of forgetting has been recent. I like to believe Robert Macfarlane when he writes, "We have forgotten ten thousand words for our landscapes, but we will make ten thousand more."

As a girl here on this land, Wren named her own ten thousand names. She experienced multitudes, infinities. She ran across the meadow in her scruffy clothes, her week-old braids, chewing on grass. "Warning: Free-Range Kid" read the headline across one such image of Wren on a poster our friend Eddie made as a joke and tacked to the bulletin board in the farm stand. Ten years ago now, Eddie started his own school out of our tool barn, which grew out of helping us homeschool Wren the year she was in fifth grade. That year, in a tiny notebook kept in the workshop, Wren scratched out a poem.

The grass is my thinker
my speaker
my twine
it winds around my toes
makes me think of my nose
oh yeah it's there
and so am I

Girlhood. So many women writers have described the idyllic sense they had as a young child in a magical earth-centric world, a world of imagination and story and heightened sensory awareness, of being a subject, not an object. And then, for most of us, it changes. Wren and Willow and I talk about this often, now that they are grown—how as women we find our infinities diminished by our culture's prescribed roles for us, and how we are trained to participate in that diminishment. How we lose our voice. In my realm as a shepherd, I have found a voice, and it insists on a particular way of doing things for the health and vitality of this land. It is the voice of the flock and the land as much as it is mine; I am the translator, which requires listening, observing, slowing down. It's easier to find voice, and to practice knowing who you are, in a realm that is inherently quiet and about listening. It's harder elsewhere. As a woman, I am still working on speaking up for what I know to be true, for what I want to defend.

But there are parallels: Just as society trains all of us to participate in some way in our own diminishment, it also trains us to contribute to the diminishment of nature. I have heard people say that participating in waste, pollution, and consumption—and turning a blind eye to so much that is toxic and violent to our earth—is simply what we have to do to live. I don't want to accept that. Our participation in the diminishment of nature is another form of self-betrayal; I want to give nature as much priority as I give myself, want to feel it as an extension of the same body that walks in my two boots over the ground. *The grass is my thinker, my speaker, my twine.*

I NEVER THOUGHT, AFTER LEAVING MY CHILDHOOD FARM, that I would ever find another place where I felt so at home, so connected to myself. I feared that the power and wholeness I felt in girlhood and lost as a teen was something I would always mourn. In shape-shifting from that deeply rooted girl, I tested my strength as a wanderer, while always longing for the place that I credited for my wholeness.

Then, I came here. Over the years, living and moving with sheep over this hillside, tending and attending to this place, I thought I had simply replaced one attachment to place with another. But in writing my love song to this hillside I have begun to trace how belonging happens, and I have begun to understand how it's not attached to particulars, not about finding any one place, one person, one thing that holds us. It is not outside of us at all. Belonging is more about the ritual and dedicated attention *within us* to something beloved that matters.

The word *belonging* encompasses longing. In nature, I feel a belonging with and a belonging to, a kinship, a reciprocity as well as a longing for greater intimacy and knowledge, to be radically present and to be loved: a be/longing in and for the greater world. But this feeling can happen anywhere; I know this now. We are participants in our own sense of belonging. It doesn't happen automatically, and has less to do with the place itself than our energy to receive it. We make it happen—with a mother, a daughter, a hillside, a lifework, a way of being—through a deep commitment over time, through our curiosity, through our merging. Belonging is a two-way embrace. It begins the moment someplace or someone says to you, "Welcome," and you receive the gift. You set your bags down inside a door, inside a heart. Then, belonging requires, more than anything, participation. It requires unpacking your bags and discarding everything you were carrying that you no longer need to fully embrace where you are. Finding belonging requires a softening of the boundaries of self. A quest to find "you are enough" becomes, all at once, "you are infinite."

To practice belonging is an act of resistance against diminishment—diminishment of ourselves and of the more-than-human world that offers all of us a home.

These are fragments I sort, turn over, rearrange. From my writing table I lift my eyes from the page often, to keep them keen to reading the script of faraway fields and trees, movement and shadow, shimmer and change of mood. My window looks out on our small pond. Around its smooth lid of ice are banks of deep snow pockmarked and slumped by yesterday's sun. The season is shifting. I hear the chickadees' spring song—one long high note, two low.

In the white oval of the pond is a perfect darker rectangle where Wren and her boyfriend, Jed, cut a hole in the ice two nights ago for a swim. In the dark, they shoveled the snow from the ice, took a hatchet and chopped into the deep black water, shed their clothes at the pond's snow-crusted edge. Wren jumped through and came back, her skin glowing and steaming, to wrap herself in a towel. The hole is sealed again, but the scar is there—scar, stamp, door, portal to another world. I look out at that square and know it goes deeper, goes where I can't follow, where I will stare at an empty doorway and make prayers.

In May, Wren will pull out of the farm road on the bicycle that she built for herself and ride all the way to Alaska. Unsupported, everything essential strapped to her bike. Two months across the continent, averaging eighty miles a day, then a week camping on the deck of the ferry that runs up the Inside Passage to Glacier Bay. She will be greeted by our friends Hank, Anya, and Linnea, her second family. Hank came to one of our gatherings for conservationists at the farm when Wren was a baby and became one of our dearest friends.

Whenever adults question the safety and sanity of Wren's plan, suggesting that she do a different adventure—Why not go with a group or on a rail trail, or at least west to east so that you aren't battling the wind?—Wren gets a stubborn look on her face, nods politely, says nothing.

I worry about her safety too. I wonder if it's irresponsible of Peter and me to let her go on this adventure alone. The more we push back, the most stubborn she becomes. Even if she wanted one of us to go, I'm certain we could not keep up. So we are building an expedition plan together, picking the best route, looping in friends to stay with along the way. Through our network of all the people who have stayed at the farm over the years, we have a solid crumb trail of loving hosts along her route west.

"I don't care if parts of it are monotonous or really hard. That's sort of the point. It's very important to me to go from the place I love most to the other place I love most," Wren told me. The Lentfers' homestead in Glacier Bay became like another home for her during her homeschool years, when we did exchanges, hosting each other's daughters.

I recognize this instinct in her. For me, there would be nothing better than long days in the mountains, walking a flock to a new place, the bleating and whistling and shouting, the dogs working the edges and everyone testing their endurance. The feeling of coming down from a thin place, the wind dropping, the piercing call of hawks replaced by sparrows chattering in the hedges, from the wide open into lush meadows and the enclosure of walls and gates, from trods to trails, trails to farm roads, farm roads to barnyards. To arrival.

Wren will go alone over four thousand miles from one place of belonging to another. She would not say it this way (and now she is rolling her eyes), but she will go through a portal, open herself to longing, and connect her worlds. On her right foot, she gave herself a stick-and-poke tattoo of a drawing she made of three bird-women with wings and fierce faces—swallows, I like to think, giving her safe passage.

Wren, who is so utterly of this place where she was born, is finding her wings. We protect a sense of wholeness and belonging for our children and then help them go, knowing that their evolution is in finding it for themselves. This feels like mother work: not just nurturing but knowing when to stand aside. Like the ewes in the field whose late-summer lambs come back to them to nurse, and they butt them away.

My mother there in Plainfield, all those years. My mother there alone after we left, tending that land. My mother who sanitized the milking machines twice a day in the blue claw-foot tub, covered in milk, covered in rain. Who cooked and baked and lit the fires. Carrying out the ashes. Carrying out the rinds. I hold grief for her loneliness, for the years I didn't go back and help her, for the times I visited and kissed her too quickly goodbye.

To tend a baby, to be a mother, can be a lonely, interior thing, especially in my culture in this time, even when you are supported by community and have a loving partner. This was something that surprised me and that I hear other women say. To raise that baby up to a grown child who will leave is another kind of loneliness that I believe all mothers everywhere understand. It's a closure but also a portal of your own. You dry your hands. You lift your gaze. The bowl of perception that has been so brimming, so overflowing with all that life requires of you, begins to empty again, inviting grief but also leaving room.

What would my mother tell me now, if she could speak? Would she tell me it's all okay, that she had no regrets for staying at home when we had all gone? She might tell me she was lonely, but that women understand loneliness, that even when we're not alone we can be lonely, because in the arc of history we were the ones who stayed, who swept, who tended, that this is embedded in our bones. The strange loneliness I feel now draws me to the fäbod tradition of the women going to the mountains together, to pass on knowledge to their daughters.

I am connected through the centuries to other women whose flocks gave them solace. The land, too, was a woman's solace. As her children flew, she stayed with the flock, and she and the flock tuned their movements to one another. The flock was her murmuration, her wingbeat, the ever-shifting pulse and pattern of her days.

A FEW DAYS AFTER THE BLIZZARD AND A SNAP OF COLD, THE sun comes to lay its softness in the snow, a loosening. Snow is a creature, ever-changing in character, shifting with the light and temperature, the wind and shadow and slope of the land. The morning brightness fills the house, inspiring me to sweep down the cobwebs and rub a cloth over the old wavy panes of glass. Snow and sun together in spring make "corn," a coarse sand-like snow that's full of water and moves easily under the sled as I haul hay out to the rams. As the sun sinks behind the ridge, the corn sets up, a thin crust forming where the slick of water meets the rapidly cooling air. I have to sled the hay out before sunset, or I'll be plunging through a crust, hurting my shins with every step.

Warm sunny days and cold nights in late winter make for perfect maple sugaring weather. All across the hillside the sugar shacks are lit up at night, a plume of steam coming from the roof, people pulling up their chairs and beers for a long night of stirring and watching the pan of bubbling sap until it gets to syrup stage. It feels like the breaking of winter's dormancy, not only for the trees but for us.

I can smell that it's a sugaring day before I even open the door. For me, that doesn't mean I need to slap my boots on quick to check the flow of sap, but the feeling is there. A quickening. A tremor of urgency. It means lambs are coming.

Peter loves sugaring, but since we moved here and got sheep he always said that even we aren't crazy enough to stay up all night for two reasons at once each spring, so we let the Vasseur brothers up the road tap our hillside of sugar maples. But this sunny cold time stirs our blood just like it does for an old sugar maker. I hope sugaring season will go on.

I hope, when Peter and I are too old and tired to do much, our children and perhaps our sheep long gone, we'll be up at night in sugaring weather, making steam in the cold air, stirring the clouds, delivering the stars.

THE WINTER NIGHT BECKONS, THE FIELDS GLITTERING. AS THE first stars come out, our boots crunch through the snow. It's deep, and the walk is slow going. Machias races across the crust to the east, chasing crows who jostle each other in the wind above, scolding like dark gods. Rue is way behind, her bum leg lurching. Seeing that we have sleds under our arms is enough to make her push through her old-dog pain.

As I walk behind Wren, the fatigue leaves my legs and leaves my heart. This is something I have given her—all those winter months when going to third and fourth grade made her sad, I'd meet her at the bus stop with two sleds, and we'd drop everything to tromp up and restore her lightness on this hill. Now just being out here with her rekindles my joy.

At the top of the logging road we can see all around us. We look out through curtains of bare-armed trees to the mountains, to the lights of neighbors high on the common road and the beams of car lights picking along the farm fields. Way below us we can see the pencil line of the river, bordered by cow fields behind Hadley's multigeneration farm, and the black spot where his father's house was destroyed by fire. Across from Hadley's we can see the farm-restaurant where Wren and Willow have worked in the summer, and farther north we can see the light-glow from the parking lot of a new brewery. Above the valley is the ridgeline, the maples gray amid the dark pines and the ancient granite outcroppings, brushstrokes on a scroll over which we trace our shifting lives.

We sit on our sleds on the log landing at the top of the road.

"This hill always feels like it's giving us a hug," Wren says.

Below us is the long snowy road and the sloping field, full darkness falling fast, draping its wings over the backs of sheep mounded there, near the big barn. They are too far away to see, but I see them; I imagine the ewes getting heavily to their feet now and then to lip small strands of blown hay from the drifts, turning and nesting their bodies round with lambs, chewing their cuds through the night. Together we wait for birthing time, when the cycle of the year will begin again.

    Rue yips and trembles in the trail, hopping her cold paws up and down, begging us to go so that she can herd and nip behind us. Her bright panting mouth is the one flash of color in the fading gray of the forest. Machias chucks off sideways through the woods, barking at wild turkeys who roost in the tallest maples. The turkeys take flight with a sound like someone discharging a parachute, then careen and tilt through the branches with great peril. This delights Machias, whose exploding white form soon disappears around the hill.

    Wren and I zip up our coats, pull our hats over our ears, and push off. We each have a "rocket" sled invented and tested by neighborhood sledding fanatics on this very hill. It's shaped like a grain scoop, or a miniature whitewater kayak. We kneel, sit our butts on our feet, buckle a strap tight across our thighs, and ride the sleds with the motion of our hips, down the road, catching air over the water bars, then swishing an S-turn through a fringe of trees and into the long open run of the field, steeper as it falls, dark now except for the glow of snow.

    We raise our arms and whoop as we fly into the night sky.

EPILOGUE

A SHEPHERD'S MIND

> And he told stories about the stars above, about the earth below. He told them to make the night pass and also because his heart was all reflections in which the soul of the world moved.
>
> —JEAN GIONO, *THE SERPENT OF STARS*

I LIE AWAKE FOR A LONG TIME, WILLING MY BODY TO LIFT itself out of the warm soft blankets. I've been half-awake for hours, tracking time and imagining what the lambs are doing in the dark. Are they lost and crying, stuck behind an old pallet, cold and hunched, being born? All night it's as if I'm watching them, a particle of my physical self—like a wisp of cloud—suspended above the places where the sheep are moving in the dark.

I step out of bed, go softly down the stairs in my nightshirt, too tired to find clothes, too tired to find socks to cushion my sore feet in my boots. Night checks are a routine, for about a month during lambing, that grind me down by degrees—similar to, but not as sustained as, having a newborn child. Machias follows me for the night watch, but Rue eyes me from her bed by the fire without lifting her head. It's 2:14 by the oven clock.

I pull my coveralls, rubber boots, two coats, and gloves over my bare skin and head out into the night. A glow from the village reflects off a bank of thick white cloud like a goose wing stretched out across the valley, its primary feathers reaching into the far hills and fading to dark gray, laced with rain.

Despite the cold drizzle, none of the sheep are sleeping in the shed. I can see them lying like stones dispersed in the field, so I head out with the flashlight, inspecting. Each ewe has her lambs tucked behind her away from the cold wind. A few lambs are asleep on their mother's broad woolly back, curled tightly like cats. By now I can identify which lambs belong to each mother and can check each one; I prowl slowly among them so they won't get up and begin to mix together. A lamb can get lost at night, and the youngest ones can't go long without nursing before they are chilled.

One brown ewe, Russet, has such a strong maternal instinct that she will steal other lambs before she has had her own. If another ewe has given birth to twins and is preoccupied with her firstborn, Russett will begin to lick and nicker to the second lamb until it bonds to her instead of to its birth mother. The first year she did this, I didn't see it happen. I came out the next morning to find Duchess—a big black ewe—with one white lamb and Russet with another: They each had a single lamb, I thought. Then two days later Russet gave birth to twins, and her larger, adopted son took all her milk and attention. I found her newborn twins in the field, bone cold and so weak that they could no longer suck. I had to feed them with a stomach tube, something I'd never done before. I managed to save one, but the other died a few hours later. It was Easter Sunday, and the congregations of two villages were in our field around a fire for the sunrise service. Everyone trooped into the farmhouse kitchen to see the two lambs in a cardboard box by the woodstove—Death and Resurrection. The third of the God trinity was pigging out in the paddock. I remember thinking it was an opportunity for a sermon on the metaphor of the flock, how it is both one and not one, each birth a distinct life and also dependent on a more complex intertwining.

I finish walking around the field; all lambs are accounted for. As I turn to go back to the barn, I hear a low guttural cry in the dark, over and over. It's too deep for a lamb's voice, and it's not the sound I hear mothers make when they are calling to find their young. It's coming from behind the long-abandoned John Deere manure spreader that my father bought at auction when I was a child. I shine the torch behind the chassis and see a young brown ewe, Bell, nuzzling a mound of earth. She's pawing and moaning. The mound of earth is a lamb, covered all over with fine grit and as cold as a salamander crawling out of the ground. The lamb's body is huge and heavy, finely made. Beside it is a mound of placenta, fibrous and red. Bell must have had a long hard labor here in her safe hideout between the spreader and the fence. Perhaps she was too tired to stand when she finally pushed the huge lamb onto the bare ground, and the lamb had some of the birth sac mixed with fine dirt clinging across its lips. In those few seconds when it left its watery womb and needed that first vital gulp of air, it would have suffocated. Now Bell has pawed and nuzzled it repeatedly, turning it over and over in the dirt until it resembles a body pulled from a grave.

When I was in college and thinking I wanted to study wildlife biology, I spent a few months in the Maasai Mara wildlife reserve of Kenya with biologists studying topi. One night poachers came and killed a bull elephant for his ivory, and the next day and for many days following there was a congregation of the mother elephant clan on the site of the killing. The mothers, grandmothers, and baby elephants went to the site day after day and rubbed their great heads on the ground, moaning, stamping, swaying. Even now when I think of it, thirty years later, I get a trembling in my chest.

I lift Bell's lamb and carry it through the dark. She follows me. I take her close to the barn where Yampa is sleeping with her tiny triplets, born yesterday. I offer Bell some water and hay, but she is too distressed to eat or drink and leaves again, pacing back to the place she gave birth. I follow her and try again to get her to come toward the barn. This time I also pick up her placenta, which hangs in a moist and shiny veil from my hand. At the barn I wrap Bell's placenta around the smallest of Yampa's newborn triplets, a wisp of a thing with a cupped face like a Chihuahua. If I can get Bell to accept this lamb, she will have a baby to care for and to feed with her abundant milk, and Yampa will have a better chance of raising the other two.

The first imprinting between mother and baby is smell, and then voice. A traditional practice among shepherds—if a ewe lost her lamb—was to skin the fleece from the dead lamb, cutting it precisely around the neck and the four wrists, opening a cut along the belly, and then peeling it off like a sweater. The orphan, dressed in this sweater like a copy of its own, might trick the mother into accepting the little pauper as her prince. This is called "grafting a lamb." If the ewe has just given birth and her hormones are still surging, that imprint of her lamb's smell is still strong, and it might work. The more time that passes for either of them, the less likely the trickery takes hold.

Yampa is very upset that I'm taking one of her three lambs, and Bell is too distressed to allow this imposter to suckle. She sniffs it but keeps turning circles as I hold the lamb to her teat. She knows this is all wrong. Both mothers glare at me. I decide to let Yampa have her triplets. I tie Bell with a halter to the gate and milk her colostrum into a jar. It's the color of dandelion and as thick as a milkshake. I get about four ounces before she's ready to murder me, but enough to share with any lamb who needs an extra boost. When I untie her she goes straight back to her mourning place behind the manure spreader.

WE HAVE THREE WARM, GLORIOUS DAYS AND NIGHTS, NO rain. Wren helps me finish pruning the apple trees and the blueberry field, and we make massive heaps of branches to burn, our arms bare in the sun. Every day the green in the fields is brighter, and the maples on the hills blush pink. Two sassy yearling ewes, Woodstock and Ada, figure out how to sneak through the fence to nibble the shoots on the lawn near the pond. The rest of the flock paces and cries. I give them extra salt mixed with dark-green sea kelp and watch them eagerly crunch the crystals between their front teeth and lick their lips. While the mothers feed, the lambs race from one end of the paddock to the other, up the manure pile and down, swirling like starlings over a lake in the dusk.

The sudden heat wave so early in spring makes all the tulips open at the flower farm down the road where our friend Avery works, and eight thousand dollars' worth of tulips are about to go to waste. I help Avery and Emily, the owner, distribute them to the food pantry and nursing home. Emily makes a gorgeous display along the steps of the library in the center of town.

The night after we distribute the flowers, the skies are full of stars. Such a loss is hard for a small farm, and I feel for Emily. The weather—and in its aggregate, the climate—has always been a farmer's most difficult and needy child, but now that child has gone on a bender, totally unpredictable, prone to extremes. We cannot sever our ties, just keep trying to adapt and take the losses. I'm not sure farming is any harder than it ever was, but it is definitely less constant and settled. No one I know expects it to be a way of life for generations anymore, the way it always has been.

The next day, a cold wind picks up, and a few flakes of snow squeeze down from a miserly gray sky. Even the trees seem to turn their hunched backs on the hillside where a day before I could feel them stretching and opening their arms.

On my night check, I turn to go back inside when I notice Yampa's tiniest lamb, hunched and staring. Had it been nursing? I have been too busy to notice. I put it under a heat lamp, and its two siblings curl up with it, but by dawn it's splayed out on the ground, no longer shivering, not lifting its head. A lamb that's curled and shivering is chilled; one that is limp and on its side is hypothermic, too cold and weak even to shiver. Its temperature, normally 102, is 99.

I put the lamb in my down coat and fold the coat over its small body, just the head with closed eyes visible. Its breathing is labored, chug, chug, chug, like the puffs of a small motor before it cuts out. Just looking at it makes me feel exhausted. I sit under the heat lamp in the corner by the cement foundation. Peter finds me there. I haven't seen him since the day before, and I feel like days have passed.

"Why don't you come inside?" he says. "I made you some eggs." Then he has to go.

I carry the lamb inside, still wrapped in down, and stand by the sink for a while, wondering where to put it. I'll need to light the woodstove. The lamb is small enough to tuck into the crook of one arm. Suddenly I'm ravenous, and I stand at the stove and eat the omelet with my fingers out of the pan—half-moon, the cheese inside still melted and the edges of bright-yellow egg lacy and shining with salted butter that dribbles down my wrist. I put the lamb in a round basket on the floor, zip my puffy coat around her, and go back out to get firewood. She lies in a limp curl, her white topknot just visible.

Once I have the fire going I notice that I've brought in lots of hay with the lamb, not to mention traces of dung and urine and other goopy substances. I sweep the floor and collect a large mound of chaff, wood dust, dirt, several dried leaves, a raisin, a few dead ladybugs, and a clump of cocklebur bound up with dog hair and saliva. I tip it all into the compost bucket to empty later in the garden. This is the daily measure of the outside coming in, and the daily task of returning what we discern as dirt to its rightful place as soil.

I have a few hours before I need to tube-feed the lamb again, and before I'm too tired to function. I head to the grocery store for a few things and to KC's Bagel Café for a bag of free day-olds; I get the last one. At a table near the register, two young women are talking; one has a very small baby in a car seat and is talking about going away on a trip, leaving him for the first time with her parents.

"I can't imagine leaving him even for a second," she says. I can hear the nervousness in her voice, and the friend is reassuring her. I want to join this conversation, then remember I am eavesdropping.

My last stop is the bike shop, to get Wren a mirror and a blinking red rear light for her bike trip across the country, safety features she has thought are unnecessary to spend money on. I'll sleep a little better if she has them, and she agrees to it for me. She leaves in less than a month. She is all grown now, but she is still my littlest lamb.

A FEW DAYS LATER, ON THE FIRST WARM EVENING FULL OF rain, the amphibians come out. The badger-eyed wood frogs and miniature spring peepers with translucent golden skin emerge from mud and leaf litter, from deep fissures of tree roots and the soft underbelly of rotten logs. They shake off winter's torpor, come out of the forest, and make their way downhill to vernal pools and open spring creeks and to our small farm pond. The amphibians come to breed, lining the pond's edges with billows of lacy frog eggs and long skeins of salamander eggs like beaded necklaces of shining onyx.

I have called Peter out to the barn to help me with a ewe in labor. She's been straining all evening and making little progress. Toni, a small silver ewe, is a first-time mother, wild with fear, her head hanging and sides heaving. Her fleece, wet from the rain, steams with her heat. Rain slaps on the metal shed roof and puddles in the trampled straw and muck by the open doorway. The other ewes, lambs nestled against them, lie all around us, chewing their cuds, billows of breath rising, their warm, wet bodies wafting lanolin's waxy smell.

I get into the pen and kneel down beside Toni, and Peter leans over the barricade to hold her head still. I take my time. Keep all my movements slow and deliberate. I look over at Peter and know he has noticed what I notice at the same moment: the sound of the first wood frogs singing in the pond. A tentative voice, then silence. A few voices answer in the night—*crrrrook*—then quiet, then again with a few more notes, like a thin, ragged heartbeat gaining strength. Wild things rejoining and rejoicing in the liquid season of spring.

As we work together in the dark, quiet, concentrating—I am by Toni's tail coaxing her lamb into position in the birth canal, and Peter is at her head, stroking one velvet cheek and whispering—we begin to notice other things. A plop, plop. A quick movement at the edge of sight. A moist and shiny thing tangled in the ewe's fleece like a jellyfish marooned in grass. An alien eye. A warbler-yellow throat, membrane-thin, pulsing with a watery song. They are among us, these small slimy amphibious gods who are the breakers of winter. We mark their arrival each year on the calendar like a birthday, marking another cycle of the earth. Their song—which grows in strength until it pulsates in the night air, ecstatic and electric—marks our final release, and the land's, from the tension and deprivation of winter.

I have pulled the lamb's legs around into the ready position to be born. Peter and I let the ewe rest in her labor for a time, and we try to catch all the frogs we can find, flopping them into the water bucket where several frogs have already fallen on their blind migratory urge from forest to pond. They are hopping over the sheep's backs, struggling like wet-suited surfers in deep sand through the damp wool. Some of the frogs, having gotten through the sheep obstacle course, bump against the wooden boards at the back of the shed, their urgent trajectory inadvertently blocked by the barn's border wall. The smallest frogs, the spring peepers, are the size of our pinky fingers and impossible to scoop up; the wood frogs are the size of small bats, with sticky webbed toes like miniature wings. There are dozens of frogs in the sheep shed with us, if not hundreds.

I am reminded, as I often am, how nature is invisible to us until we start to notice, and how this noticing is both a source of wonder and undoing, of transcendence and grief. It's a kind of noticing akin to falling in love, or having a child; when you begin to fully see all the fantastic, alien, mind-blowing, beautiful, and generous creatures and formations of nature, you feel as if you will come apart with joy and also with fear of losing this thing that you can now feel sleeping on your chest, breathing in your face, and wrapping its fingers trustingly around yours. This is how it's been, at least for me.

A few minutes later, the young ewe is licking her newborn lamb, drunk on its long, flimsy legs, falling and getting up again and again, determined to find milk within minutes of birth. Peter and I pick up two buckets of frogs and head out into the night. We tip the roiling buckets of life into the inky dark at the edge of the pond, which has suddenly gone quiet, sensing our presence. The frogs slip away, joining others, whispering their lines for the full-throated chorus about to come as we leave them to their ritual and head home to bed.

A DREAM IN MY TWENTIES—SO VIVID THAT I REMEMBER IT still—was about being a small animal that lived under a hedge. My role in life as this hedgehog-person was to come out at night and make notes about the moon and record the exact details of the moon's passage through the sky. In my dream this felt like a monumental responsibility, the most important thing to get right. It felt as though having such a responsibility would be enough to live a good and full life.

Lately, I have other dreams, most recently a dream about standing on a jetty on the edge of a sea and watching dolphins rising and falling in the waves. As I watched, one dolphin came over to where I stood and spy-hopped out of the water, as whales do, and looked me in the face with a fierce eye. We were the same height—a newly arrived mammal species who has forever altered the world, and an ancient mammal who walked the land nearly fifty million years before us and evolved to swim in the sea—looking at each other nose-to-nose like two wrestlers before a fight. Then the Ancient One swam away.

The dream meandered, then returned to this jetty, except this time the sea had dried up. I looked out on an endless red, cracked mudflat that bore not a single sign of life. I remember, in the dream, having the thought that my species had won, and was left with nothing at all. I woke and lay there with that image so vivid in my mind and thought of Annie Dillard's advice in *The Writing Life* about not hoarding one's ideas: "Anything you do not give freely and abundantly becomes lost to you. You open your safe and find ashes." I saw humans coming back to nature, opening the closet door to all their hoarding and striving, and finding nothing there but a pile of bones.

Part of me lives in that futuristic dream, looking out over a cracked ocean of desolate nothingness, but a larger part of me is that hedgehog note keeper, that moon scribe, not wanting to miss anything—any rite of spring, any nation of unobtrusive mass in the singing grass. In writing this love song I didn't want to sing only of loss, but to sing of the immensity and wonder of what I've found, what has been generously given to me, what I hold dearly and deeply cherish.

I had no inkling—when we first unloaded eight sheep into the fields at Knoll Farm nearly twenty-five years ago—of the impact that learning to be a shepherd would have on me. I see, in the fascination visitors to our farm express over and over, that others share this ancient, almost primal pull to the shepherding life. It's not the way of life people crave so much as a way of being, I think, something they sense encloses both wandering and home, the wild and the cultivated, uncertainty and joy, solitude and also being part of a community so broad that it encloses all of you. Shepherding could be a fertile edge place, both real and metaphorical, that we might draw on now in our collective lostness, to find our way toward healing ourselves, each other, and the land. It offers us a way to care for the more-than-human, and to find a holding on, inside ourselves, as things whirl apart. But, you don't have to become a sheep farmer to cultivate shepherd's mind, which is about finding a way to listen, to tend, and to immerse in the living world. Sheep may be one way. Most importantly, may you find your own.

A SHEPHERD'S GLOSSARY

Bale: a packed unit of hay, square or round; can also refer to a unit of pressed wool

Bellwether: originally an experienced sheep given a bell to lead a flock; now mainly used figuratively for a person acting as a lead and guide

Bottle lamb: an orphan lamb reared on a bottle

Broken-mouthed: used to describe a sheep that has lost or broken its incisor teeth, usually a mark that condition will decline

Bum lamb or **bummer:** a lamb that steals milk from other mothers; often orphaned

Cast: unable to regain footing, possibly due to lying in a hollow with legs facing uphill and/or having a heavy fleece

Clip: all the wool from a flock

Crimp: the natural fine waves seen in wool fiber; usually the closer the crimps, the finer/softer the wool

Cull: a sheep no longer suitable for breeding and sold for meat

Dags: clumps of dried dung stuck to the wool of a sheep

Drench: an oral medicine, usually for treating parasites

Driving or **droving:** walking animals from one place to another

Dunged-up: the condition of an area of pasture where sheep have concentrated and trampled their dung (poop) and no longer find the grass palatable

Ewe: a female sheep capable of producing lambs

Fleece: the wool covering of a sheep, equivalent of fur; one fleece is all the wool taken off one sheep during shearing; more often called wool when it's off the animal, and yarn when it's spun into thread

Flushing: providing especially nutritious feed in the few weeks before mating to improve fertility, or in the period before birth to increase lamb birth-weight

Fly-strike: flies hatching and feeding in an open sore on an animal's body; usually leads to serious infection or death

Fold or **sheepfold:** a pen or three-sided shed in which a flock is kept overnight to keep the sheep safe from predators, or to allow the collection of dung for manure

Freemartin: an infertile intersex animal that has the external genitalia of a female but other characteristics of a male; the word comes from the tradition of fattening "free" (unbred) cows and sheep for St. Martin's Day in the fall

Gimmer: a young female sheep, usually before her first lamb; also called a theave

Grease wool: or "in the grease"; a sheep's fleece before it is washed

Gummer: a sheep so old that it has lost all of its teeth

Heft: a piece of pasture to which an animal has become attached, which it knows how to find again and has the instinct on which to stay, to be "hefted"; from an Old Norse word meaning *tradition*

Lamb: a young sheep in its first year; also the meat of younger sheep

Lanolin: a greasy substance in wool, secreted by the sheep's skin; also called wool fat, wool wax, wool grease, or yolk; protective and high in vitamin D

Micron: one millionth of a meter; a measure of fiber diameter in individual wool hairs; used to determine the fineness or softness of a fleece

Moorit: Icelandic for brown; a brown sheep

Mouflon: *Ovis gmelini*, the wild ancestor of all sheep; also a fleece pattern with a light belly and throat and a darker body

Mutton: the meat of a sheep more than fifteen months of age

Ram: an uncastrated adult male sheep; also tup

Rooing: collecting the wool fleece found on fence wires or twigs where it has naturally snagged

Second cuts: short pieces of fleece made when a shearer goes over an area already sheared a second time; not desirable as they will be too short to spin

Shearling: a rug or fabric made of a sheep's skin with fleece attached; also sheepskin or hide

Shepherd's crook: a wooden staff with a curved end used to catch sheep by the neck or leg

Shieling: a summer dwelling on a seasonal alpine pasture, usually made of stone

Skirting: sorting a fleece to remove dags, second cuts, and bits of hay and other matter

Staple: a group of wool fibers that formed a cluster, looked at often to determine the length of the fleece and suitability for spinning; many mills cannot spin wool over or under a certain "staple length"

Suint: sheep sweat and dirt, rich in potassium salts, caught in wool

Teaser ram: an infertile ram put in with ewes at breeding time to help bring them into heat for breeding

Theave or theaf: a young female sheep, usually before her first lamb; used more in England; also called a gimmer; in New England, a one-winter ewe.

Thel or þel: Properly spelled þel but pronounced "thel"; the softer, shorter inner hairs on the dual-coated fleece of many northern breeds

Tog: the longer, coarser outer guard hairs on a dual-coated fleece; sheds rain and snow

Transhumance: the seasonal movement of animals to find the best pasture according to the natural topography of the land; literally "to cross the ground"

Trods: the paths sheep make by following single file

Tup: an alternative term for ram

Tupping: the act of mating; sometimes used to describe the breeding season

Wether: a castrated male sheep often kept for its fleece or as a companion to the breeding ram, as a leader, or as a "teaser ram"

Wool break: a marked break in the fleece, producing distinct weakness in one part of the staple, usually marking a time of illness or stress

NOTES

CHAPTER 2: INITIATION

18 *Here you are crossed* Jean Giono, trans. Jody Gladding, *The Serpent of Stars* (Brooklyn, NY: Archipelago, 2004), back cover.

19 *The night we munched* Giono, *The Serpent of Stars*, 12.

25 *Basho's haiku are the record* Jane Hirshfield, "Basho as Teacher," *Tricycle*, Spring 2002, https://tricycle.org/magazine/basho-teacher/.

28 *She would see into the windows* N. Scott Momaday, *House Made of Dawn* (New York: Harper & Row, 1968), 54.

34 *a ghost is* Lauren Groff, "On Growing Up in What Felt Like the Middle of Nowhere," Lit Hub, September 28, 2021, https://lithub.com/lauren-groff-on-growing-up-in-what-felt-like-the-middle-of-nowhere.

35 *We are passing through the gate* Robert Hass, *Like Three Fair Branches from One Root Deriv'd* (New York: Ecco Press, 1974), 13–14.

37 *May I be forgiven* Giono, *The Serpent of Stars*, 12.

CHAPTER 3: THE FLOCK IS THE LAND

49 ***As microbes*** Fred Provenza, *Nourishment* (White River Junction, VT: Chelsea Green, 2018), 62. An excellent book full of revelations about the complex interactions of plants and animals, nutrition and evolution. Provenza takes the view that "organisms aren't merely adapting to environments, they are actively participating in creating relationships with them," (52) thereby constantly colearning and coevolving.

49 ***The peaceful ruminating sheep*** Don Mitchell, *The Souls of Lambs* (Boston: Houghton Mifflin, 1979), 23. A beautiful, poetic book that inspired the author's daughter, singer Anaïs Mitchell, to compose an equally beautiful song called "Shepherd."

52 ***through flavors of foods*** Fred Provenza, "Food Production Systems Involved and Evolving with Landscapes," *Nomadic Peoples*, March 2021, vol. 25, no. 1.

58 ***I saw a documentary*** I can no longer find a source for the film I saw so many years ago, but this tradition of singing to mother animals to help them produce milk is part of many traditional cultures around the world. There's a beautiful documentary shot in Mongolia, *The Story of the Weeping Camel*, that is widely available.

60 ***James Rebanks, in his beautiful book*** James Rebanks, *The Shepherd's Life: Modern Dispatches from an Ancient Landscape* (New York: Flatiron, 2015).

CHAPTER 4: SPRING MEADOW

69   *flight zone*   For those interested in more about herding and low-stress animal handling, I highly recommend the work of Temple Grandin as well as the work of Bob Kinford and others in the art of instinctual migratory grazing.

71   *the silhouettes of tiny swallows*   Henry David Thoreau's *Wild Fruits*, ed. Bradley P. Dean (New York: W. W. Norton, 2000), 257. I have always thought that birch seeds on the snow look like a flock of birds, so I was delighted to come across this passage: "Those [seeds] of this species are peculiarly interesting having the exact form of stately birds with outspread wings, especially of hawks sailing steadily over the fields, and they never fail to remind me of them when I see them under my feet."

74   *placid and self-contained*   Walt Whitman, *Leaves of Grass: The First (1855) Edition* (New York: Penguin, 1959), 55. Later in the same poem are lines I very much related to regarding my sheep: "they show their relations to me and I accept them / They bring me tokens of myself."

74   *leaf hay*   See William Bryant Logan, *Sprout Lands* (New York: W. W. Norton, 2019), for an interesting account of coppicing and pollarding trees in Scandinavia to produce winter feed for livestock. Leaf hay was a solution where grass was scarce; tannins in some leaves have also been shown to deter parasites and have nutritional benefits.

74	*smelled before they are seen* Thoreau, *Wild Fruits*, 15.
77	*The meadows observed and recorded* Verrazzano, Giovanni Da, and New York State Historian, *Verrazzano's Voyage Along the Atlantic Coast of North America* (Albany, NY: The University of the State of New York, 1916), https://www.loc.gov/item/17027055/.
77	*Our farm sits above the village* History of Waitsfield courtesy of the Waitsfield Historical Society. The early history of our farm, which overlooks the village of Waitsfield but is in Fayston, is recorded by the Fayston Historical Society in a booklet compiled by Anna Bixby Bragg: "The Early Years, Fayston, Vermont 1798–1898: Historical Address at the Centennial Celebration of the Town of Fayston, Vermont, Including Short Biographies, 1898." Rufus Barrett, who first settled our farm in 1804, was the second name on the subscription of the town, came from Connecticut and "built a frame house," and later donated land for the cemetery, parsonage, and the church in Irasville.
78	*Merino sheep are a breed* The history of Merino sheep in the United States has been written about extensively, and I drew from many sources, including: Jan Albers, *Hands on the Land: A History of the Vermont Landscape* (Cambridge, MA: MIT Press, 2000), 145–49; Steven Stoll, *Larding the Lean Earth: Soil and Society in Nineteenth-Century America* (New York: Hill and Wang, 2002), 108–15; and Howard S. Russell, *A Long, Deep Furrow: Three Centuries of Farming in*

NOTES   271

New England (Hanover, NH: University Press of
New England, 1982), 158–59, 201–11.

78 *the salt put out on stones along the way* Robert R.
Livingston, *Essay on Sheep: Their Varieties—Account
of the Merinoes of Spain, France, &c* (New York: T.
and J, Swords, 1809), 39. "They suppose in Spain
that salt contributes greatly to the fineness of the
wool. The shepherd places fifty of sixty flat stones
at about five paces from each other; he strews salt
upon, and leads the sheep among them."

79 *We will use our utmost endeavors* Journals of the
Continental Congress 1774–1789, vol. 1 (Washington:
Government Printing Office, 1904), 78.

79 *The British navy targeted sheep* L. H. Butterfield
and Marc Friedlaender, eds., *The Adams Papers,
Adams Family Correspondence*, vol. 3, April
1778–September 1780 (Cambridge, MA:
Harvard University Press, 1973), 97.

79 *I hope to attain my object* Livingston, *Essay on
Sheep*, 7.

80 *Out of 6,000, only 610 survived* Stoll, *Larding the
Lean Earth*, 111.

80 *By 1850, the state* Albers, *Hands on the Land*, 148.

81 *Judge Carpenter caught it in his arms* Anna
Bixby Bragg, "The Early Years, Fayston,
Vermont, 1798–1898," in *Historical Address at
the Centennial Celebration of the Town of Fayston,
Vermont, Including Short Biographies*, Fayston
Historical Society, 1898.

82 *Within a few years, sheep* George P. Marsh, *Man and
Nature, Or, Physical Geography as Modified by Human*

Action, ed. David Lowenthal (Seattle: University of Washington Press, 2003), footnote 24.

82 *The felling of the woods* Marsh, Man and Nature, 3.

83 *the forest waters the farm* Yearbook of the United States Department of Agriculture, 1895 (Washington, DC: Government Printing Office, 1896), 337.

84 *a minor Nation celebrates* Emily Dickinson, "Further in Summer Than the Birds," ed. R. W. Franklin, in *The Poems of Emily Dickinson* (Cambridge, MA: Harvard University Press, 1999), 895.

85 *The field was covered* Thomas Berry, "The Meadow Across the Creek," in *The Great Work: Our Way into the Future* (New York: Bell Tower, 1999), 12.

85 *We have silenced* Berry, The Great Work, 17.

CHAPTER 5: STALKING COYOTES

93 *Coyotes arrived east* An interesting biography of the coyote and history of our attempts to control populations: Dan Flores, *Coyote America: A Natural and Supernatural History* (New York: Basic, 2016). Flores's book led me to many primary sources and studies on coyote behavior and control. For eastern coyote and interbreeding with wolf see: Jonathan G. Way et al., "Genetic Characterization of Eastern 'Coyotes' in Eastern Massachusetts," *Northeastern Naturalist* 17, no. 2 (2010): 189–204. Also: Adolph Murie, *Ecology of the Coyote in Yellowstone, Fauna of the National Parks of the United States*, No. 4 (Washington, DC:

US Department of Interior, National Park Service, 1940). Murie's studies were the first that endeavored to understand coyotes in their natural habitat and to argue against the National Park Service's long tradition of predator eradication. See also: Barry Lopez, *Giving Birth to Thunder, Sleeping with His Daughter: Coyote Builds North America* (New York: Avon, 1977).

101 *I was young then, and full* Aldo Leopold, *A Sand County Almanac and Sketches Here and There* (Oxford University Press, 1949), 130.

102 *The origins, liveliness, and durability* Lewis Hyde, *Trickster Makes This World* (New York: Farrar, Straus and Giroux, 1999), 9.

104 *When the smart coyotes* Flores, *Coyote America*, 87–97.

105 *With the exception of Adolph* Murie, *Ecology of the Coyote in Yellowstone*.

105 *This letter also addresses* Robert Crabtree for Yellowstone Ecological Research Center, letter to Interested Person or Party, revised June 21, 2012, http://www.predatordefense.org/docs/coyotes _letter_Dr_Crabtree_06-21-12.pdf.

CHAPTER 6: LIGHTNING

119 *It says that when you build a thing* Christopher Alexander et al., *A Pattern Language* (New York: Oxford University Press, 1977), xiii.

122 *three rules of shepherding* James Rebanks, *The Shepherd's Life: Modern Dispatches from an Ancient Landscape* (New York: Flatiron, 2015), 207–8.

CHAPTER 7: PASSERINE

136 *In one study I came across* Amelia R. Cox et al., "Demographic Drivers of Local Population Decline in Tree Swallows in Ontario, Canada," *The Condor* 120, no. 4 (2018).

138 *A lexicon of land-love* Robert Macfarlane; original source unknown.

138 *Well it hurts to be a butterfly* Kat Wright, "Reminder," live at Knoll Farm, 2021, https://youtu.be/YqFRqPF-_io.

139 *It is much more probable* Gilbert White, *The Natural History of Selborne* (London: Oxford University Press, 1789), 117. White's letter XVIII goes on at length with his observations and delight in swallows. See: Letter XII (47–48) for his musings on whether or not swallows overwinter in the mud under the river. See also: David Quammen, "The Swallow That Hibernates Underwater," in *Wild Thoughts from Wild Places* (New York: Scribner, 1998).

140 *As summer weareth out* P. G. Foote, ed., *Olaus Magnus: Description of the Northern Peoples*, trans. Peter Fisher and Humphrey Higgens, vol. 3 (London: Hakluyt Society, 1996), 980–81.

INTERLUDE II

148 *Read a lot* Barry Lopez, *About This Life* (New York: Vintage, 1999), 14.

148 *Heavenly Day* Patricia J. Griffin, "Heavenly Day," https://lyrics.lyricfind.com/lyrics/patty-griffin-heavenly-day.

CHAPTER 8: SHEARING DAY

156 ***forces the sheep into a position*** Robert R. Livingston, *Essay on Sheep: Their Varieties—Account of the Merinoes of Spain, France, &c* (New York: T. and J. Swords, 1809), 94.

164 ***The first woolen factory*** Deb Fuller and Ann Buermann Wass, "'Without Being Obliged to Send 3000 Miles for the Cloth': The American Wool Industry, 1789–1815," Textile Society of America Symposium Proceedings, 2012, 686, https://digitalcommons.unl.edu/tsaconf/686.

164 ***We anxiously look forward*** *Raleigh Register, and North-Carolina Weekly Advertiser*, May 25, 1809.

164 ***after laying them on one side*** Livingston, *Essay on Sheep*, 91–96.

165 ***fibershed*** Rebecca Burgess is one person at the forefront of the effort to reenergize local supply chains for the growing and manufacturing of natural fiber-based cloth. See: Rebecca Burgess, *Fibershed: Growing a Movement of Farmers, Fashion Activists, and Makers for a New Textile Economy* (White River Junction, VT: Chelsea Green Publishing, 2019). See also: books by Clara Parkes.

165 ***Whether it's shepherding or shearing*** Clara Parkes, *Vanishing Fleece: Adventures in American Wool* (New York: Abrams, 2019), 169.

CHAPTER 9: GIFTS

177 ***We were as twinn'd lambs*** William Shakespeare, "The Winter's Tale," in *Complete Works of William Shakespeare*, vol. 2 (New York: A. L. Burt, 1900), 67.

198   *knowledge of objective truth* William Bryant Logan, *Sprout Lands* (New York: W. W. Norton, 2019).

CHAPTER 10: TRACES

199   *It is not thinking* William Bryant Logan, *Sprout Lands* (New York: W. W. Norton, 2019), 108.

CHAPTER 11: DAMAGE AND HEALING

212   *About 1,400 years ago* All sheep have a single common ancestor in the Asiatic mouflon, but Icelandic sheep and the Navajo-Churro—though evolving in very different landscapes—both retain many of the ancient breed characteristics, such as short tails and a dual-coated fleece. For more explanation see: Doreen M. Gunkel, "The Genetic and Historical Linkage Between the Old Norwegian Sheep, the Icelandic Sheep, and the Navajo Churro," Medievalists.net, May 11, 2012, https://www.medievalists.net/wp-content/uploads/2012/05/12051901.pdf.

213   *According to the Diné* An in-depth history of the Navajo-Churro sheep and their cultural significance to the Diné, from which I drew most of my material: Marsha Weisiger, *Dreaming of Sheep in Navajo Country* (Seattle: University of Washington Press, 2009). See also: The Navajo Sheep Project, https://www.navajosheepproject.org.

215   *The flock-reduction years* William Cronon, *Dreaming of Sheep*, 229.

NOTES

217 ***That these consequences were unanticipated*** Cronon's phrase, "to think with sheep," is borrowed from Claude Levi-Strauss's notion that plants, animals, and objects are *"bonnes a penser,"* in the sense that we can understand ourselves only within a web of complex interrelationships, symbols, associations, and meanings cocreated with the world around us. William Cronon, preface in *Dreaming of Sheep*, xii.

INTERLUDE III

224 ***Because the occupation of the masters*** Jean Giono, trans. Jody Gladding, *The Serpent of Stars* (Brooklyn, NY: Archipelago, 2004), 54–62.

226 ***'My boy,' he said*** Giono, *The Serpent of Stars*, 34.

CHAPTER 12: WINTER

233 ***the poor mans pick-purse*** *Kalendar of Shepherds: Being Devices for the Twelve Months* (England: Sidgwick & Jackson, 1908), 16, https://wellcomecollection.org/works/f4824s6t.

234 ***the windes begin to knocke*** *Kalendar of Shepherds*, 44.

237 ***High in the summer farms*** Jesper Larsson defines a summer farm (*fäbod*) as "a periodic summer settlement for the purpose of using common pastures for grazing and processing milk into non-perishable products. The summer farm was a specialised feminine workplace and a function within the farm and agriculture . . . The

transformation of herding from a male task to an occupation for women started in medieval times and culminated in the seventeenth century." Over many centuries, the practice of summer farming evolved, and no doubt I have taken some liberties in compressing the elements and describing them as I did. Jesper Larsson, "The Expansion and Decline of a Transhumance System in Sweden, 1550–1920," *Historia Agraria*, April 2012, 11–39. In *Sprout Lands*, William Bryant Logan interviews farmers in Western Norway about the traditional transhumance to summer farms and imagines the revelry and the cheeses they ate. William Bryant Logan, *Sprout Lands* (New York: W. W. Norton, 2019), 188.

239   ***The musical structure***   Anna Ivarsdotter, "And the Cattle Follow Her, for They Know Her Voice," in *Pecus: Man and Animal in Antiquity* (Rome: The Swedish Institute in Rome, 2004), 150–53.

240   ***We have forgotten ten thousand words***   Robert Macfarlane, *Landmarks* (Penguin, 2015), 14.

EPILOGUE: A SHEPHERD'S MIND

261   ***Anything you do not give***   Annie Dillard, *The Writing Life* (Harper Perennial, 1989), 79.

BIBLIOGRAPHY

Abram, David. *Becoming Animal: An Earthly Cosmology.* Pantheon, 2010.

Bashō, Matsuo. *The Narrow Road to the Deep North and Other Travel Sketches.* Penguin, 1966.

Berry, Wendell. *The Art of the Commonplace.* Counterpoint, 2002.

Berry, Wendell. *The Unsettling of America: Culture and Agriculture.* Sierra Club, 1998.

Brown, Lauren. *Grasses: An Identification Guide.* Roger Tory Peterson Institute, 1979.

Burgess, Rebecca. *Fibershed: Growing a Movement of Farmers, Fashion Activists, and Makers for a New Textile Economy.* Chelsea Green Publishing, 2019.

Chaskey, Scott. *Soil and Spirit: Cultivation and Kinship in the Web of Life.* Milkweed Editions, 2023.

Coulthard, Sally. *A Brief History of the World According to Sheep.* Anima, 2020.

Cronon, William. *Changes in the Land: Indians, Colonists, and the Ecology of New England.* Hill and Wang, 1983.

Dethier, Vincent G. *Crickets and Katydids, Concerts and Solos.* Harvard University Press, 1992.

Dillard, Annie. *Pilgrim at Tinker Creek.* Harper and Row, 1974.

Dillard, Annie. *The Writing Life.* Harper Perennial, 1989.

Flores, Dan. *Coyote America: A Natural and Supernatural History.* Basic, 2016.

Giono, Jean. *The Serpent of Stars.* Translated by Jody Gladding. Archipelago, 2004.

hooks, bell. *Belonging: A Culture of Place.* Routledge, 2009.

Hyde, Lewis. *Trickster Makes This World: Mischief, Myth, and Art.* Farrar, Straus and Giroux, 1998.

Leopold, Aldo. *A Sand County Almanac: And Sketches Here and There*. Oxford University Press, 1966.

Logan, William Bryant. *Sprout Lands: Tending the Endless Gift of Trees*. W. W. Norton, 2019.

Lopez, Barry. *About This Life: Journeys on the Threshold of Memory*. Knopf, 1998.

Lopez, Barry. *Crossing Open Ground*. Scribner, 1988.

Lopez, Barry. *Giving Birth to Thunder, Sleeping with His Daughter: Coyote Builds North America*. Avon, 1977.

Kessler, Brad. *Goat Song: A Seasonal Life, A Short History of Herding, and The Art of Making Cheese*. Scribner, 2009.

Kimmerer, Robin Wall. *Braiding Sweetgrass: Indigenous Wisdom, Scientific Knowledge, and the Teachings of Plants*. Milkweed Editions, 2020.

Livingston, Robert R. *Essay on Sheep*. T. and J. Swords, 1809.

Macfarlane, Robert. *Landmarks*. Penguin, 2015.

Mitchell, Don. *The Souls of Lambs: A Fable*. Houghton Mifflin, 1979.

Moerman, Daniel. *Native American Medicinal Plants: An Ethnobotanical Dictionary*. Timber, 2009.

Momaday, N. Scott. *House Made of Dawn*. Harper & Row, 1968.

Parkes, Clara. *Vanishing Fleece: Adventures in American Wool*. Abrams, 2019.

Provenza, Fred. *Nourishment: What Animals Can Teach Us About Rediscovering Our Nutritional Wisdom*. Chelsea Green, 2018.

Provenza, Fred and Michel Meuret, editors. *The Art & Science of Shepherding: Tapping the Wisdom of French Herders*. Acres USA, 2014.

Rebanks, James. *Pastoral Song: A Farmer's Journey*. Mariner, 2021.

Rebanks, James. *The Shepherd's Life: Modern Dispatches from an Ancient Landscape*. Flatiron, 2015.

Salmón, Enrique. *Eating the Landscape: American Indian Stories of Food, Identity, and Resilience*. University of Arizona Press, 2012.

Savory, Allan, with Jody Butterfield. *Holistic Management: A New Framework for Decision Making.* Island Press, 1991.

Stoll, Steven. *Larding the Lean Earth: Soil and Society in Nineteenth-Century America.* Hill and Wang, 2002.

Thoreau, Henry David. *Wild Fruits: Thoreau's Rediscovered Last Manuscript.* Edited by Bradley P. Dean. W. W. Norton, 2002.

Weisiger, Marsha. *Dreaming of Sheep in Navajo Country.* University of Washington Press, 2009.

White, Gilbert. *The Natural History of Selborne.* The Folio Society, 1962.

ACKNOWLEDGMENTS

Because I came back to writing and reading after years of being immersed in farming, I am immensely grateful to my seedkeepers, who believed in this project even before its germination. Those people include Peter Forbes, Peter Whybrow, Rani Arbo, Scott Kessel, Kerry Brady, Lyedie Geer, Jasmine Sudarkasa, Lindsay Jernigan, Dan Brayton, Megan Mayhew-Bergman, Scott Chaskey, Rowan Jacobsen, Peter Lourie, Anni Mackay, Scott Russell Sanders, Olivia Hoblitzelle, Danyelle O'Hara, Nick Triolo, and many others (forgive me and know I carry dearly your contribution). So many writers pointed the way for me, their words and thoughts guiding me through the writing of this story and these sentences, and more than that, guiding me to better understand this life: Robert Macfarlane, Barry Lopez, Michael Ondaatje, Henry David Thoreau, Antoine de Saint-Exupéry, Jean Giono, Jody Gladding, Annie Proulx, bell hooks, Bayo Akomolafe, Mary Austin, Annie Dillard, James Rebanks, J. Drew Lanham, and many others. To the many musicians whose lyrics and sentences made mine that much better, especially Patty Griffin, Rani Arbo, and Ben Cosgrove, thank you.

I am filled with gratitude for all the people who stepped in to protect my time, who gave their precise encouragement and improvements, and who helped to midwife this long labor and bring it into the world. To my readers who took valuable time out of their lives to give me invaluable comments and faith in my manuscript, I cannot thank you enough: Rani Arbo; Hank Lentfer; Rowan Jacobsen; Ellen Wayland-Smith; my father, Peter Whybrow; my daughter, Wren Fortunoff; and my partner, Peter Forbes. Writing can be a solitary endeavor . . . it has to be. There were many winter days and weeks when I would head to my writing desk very early, in my pajamas, and not dare to check email or answer the phone or even eat or talk to anyone before noon so as not to lose the voice in my head; but still, this was a communal blossoming. People who do not even know that I have written a book have made it possible by helping create the stories I wanted to tell.

I am grateful to my agent, Paul Bresnick, for his generous encouragement and support all along and his faith in my decisions. I could not be more fortunate to have Daniel Slager as my editor, who made the book so much better with his gentle but astute suggestions and whose instinct about words and books I trust completely. It's been an amazing partnership, and my life in books would be much less rich without you, Daniel. I could not be more grateful to Lauren Langston Klein, Theresa Cameron, and Benjamin McKinney for their expert shepherding of the manuscript; to Mary Austin Speaker, for the gorgeous cover; and to the whole amazing team at Milkweed Editions, for their artistry, devotion, skill, and kindness along the way.

I'd like to thank the lands that gave me new horizons away from the farm when I needed to get my thoughts straight: The Mesa Refuge, where I wrote the first twenty-five pages, none of which I used but that somehow informed what came later, and where the beaches of Point Reyes once again grabbed my heart; the land at Heron's Rip in fall and again in spring, bleak and raining, the tides pulsing in and out of the rip, perfect for staying indoors and knuckling through the fifth draft; the house

ACKNOWLEDGMENTS

at Blow-Me-Down, where I was able to have deep, silent conversations with my mother; the small study in my father's Los Angeles apartment where I worked while he was getting cancer treatment, having long, illuminating conversations with him about books and writing, despite the little energy he had. Thank you, too, to all the people who cared for the farm on writing days over the last two years: Peter, Wren, Rafa, Avery, Maddie, Kyla, Zoe, Cale, Liraz, and Sophia.

For my family, I have love and gratitude beyond measure. To Peter, my longest and deepest writing partner, whose insights made this book stronger and who has been tireless at creating a meaningful life. To Wren, well, I don't have the words. Your art in these pages makes me so proud and happy. Thank you for being my copilot in countless ways. Thank you to Willow for all your support and for being an inspiration in how to have a passion and find one's way in the world. Thank you to my father, Peter, for being such an intuitive reader of my words and for shaping so much of my thinking and love of books. Thank you to my mother, Ruth, who also shaped that love of books and of nature, and who gave so much of herself and her heart to me and to the world. It's especially poignant to have written these memories of her, as she died while this book was being published. Thank you to my sister, Kate, for being a steady friend, support, and partner in hard work and parent-caring all these years.

Finally, to the sheep, to the births of nearly one thousand lambs whose lives are embedded in these pages. To the flock and to the land and all the creatures of the land. I love you. This is your song.

HELEN WHYBROW is the author of *A Man Apart: Bill Coperthwaite's Radical Experiment in Living* and *Dead Reckoning: Great Adventure Writing from 1800–1900*. She is also the editor of many anthologies, including *Hearth: A Global Conversation on Community, Identity, and Place* and *Coming to Land in a Troubled World*. Her writing has appeared in *Cagibi*, *Hunger Mountain*, *EatingWell*, and *Orion*. She is a visiting professor at Middlebury College and has taught at the Bread Loaf Environmental Writers' Conference. She lives in the Green Mountains of Vermont, where she shepherds a two-hundred-acre organic farm.

WREN FORTUNOFF grew up farming and loves very long trail runs in the mountains. She has illustrated on paper bags, T-shirts, dead trees, feet, and walls. This is her first book.

milkweed
EDITIONS

Founded as a nonprofit organization in 1980, Milkweed Editions is an independent publisher. Our mission is to identify, nurture, and publish transformative literature and build an engaged community around it.

We are based in Bdé Óta Othúŋwe (Minneapolis) in Mní Sota Makhóčhe (Minnesota), the traditional homeland of the Dakhóta and Anishinaabe (Ojibwe) people and current home to many thousands of Dakhóta, Ojibwe, and other Indigenous people, including four federally recognized Dakhóta nations and seven federally recognized Ojibwe nations.

We believe all flourishing is mutual, and we envision a future in which all can thrive. Realizing such a vision requires reflection on historical legacies and engagement with current realities. We humbly encourage readers to do the same.

milkweed.org

Milkweed Editions, an independent nonprofit literary publisher, gratefully acknowledges sustaining support from our board of directors, the McKnight Foundation, the National Endowment for the Arts, and many generous contributions from foundations, corporations, and thousands of individuals—our readers. This activity is made possible by the voters of Minnesota through a Minnesota State Arts Board Operating Support grant, thanks to a legislative appropriation from the Arts and Cultural Heritage Fund.

Our work is supported by generous individuals who make a gift to underwrite books on our list. Special underwriting support for *The Salt Stones* was provided by Paulita LaPlante.

Interior design by Mike Corrao
Typeset in Hightower

Hightower was originally designed in 1994 by Tobias Frere-Jones for the American Institute of Graphic Arts. Its design is loosely based on the stylings of Venetian engraver Nicolas Jenson and his printings from the 1470s.